THE GREENWAY IMPERATIVE

The
GREENWAY IMPERATIVE

Connecting Communities and Landscapes for a Sustainable Future

CHARLES A. FLINK

Foreword by Keith Laughlin

University of Florida Press

Gainesville

Frontispiece: A lone cyclist on the Razorback Regional Greenway south of Springdale, Arkansas. Photo by Nancy Pierce/rights holder Alta Planning + Design.

25 24 23 22 21 20   6 5 4 3 2 1

Library of Congress Control Number: 2019943728
ISBN 978-1-68340-115-5

UF PRESS

UNIVERSITY
OF FLORIDA

University of Florida Press
2046 NE Waldo Road
Suite 2100
Gainesville, FL 32609
http://upress.ufl.edu

CONTENTS

FOREWORD

I first became aware of Chuck Flink's work in 2001, not long after I became president of Rails-to-Trails Conservancy. The occasion was the publication of *Trails for the 21st Century*, of which Chuck was a coauthor. As I was new to the world of trails and greenways, Chuck's book was part of my professional orientation. I met Chuck in the flesh soon thereafter, and he has been a valued colleague ever since.

The dictionary defines imperative as "an essential or urgent thing." When I first joined this movement, most public officials did not consider greenways to be essential or urgently needed. Rather, they viewed them as optional amenities that were nice to have, but not essential. All of that has changed in the last two decades. Today, any forward-looking local leader understands that greenway systems are an essential element of a healthy twenty-first-century community. In a word, greenways have become *imperative*.

Why the change? Why have trails and greenways become "imperative"?

I suggest two related reasons. First, due to a surge in greenway development in recent decades, there are now thousands of miles of greenways traversing the American landscape that are enjoyed by tens of millions of Americans every year. Greenways are wildly popular, and public officials know it. Second, as greenways have become ubiquitous and their use has become a regular part of life for millions of Americans, their multiple benefits have become more widely understood and appreciated.

These numerous benefits can best be summarized as components of *health* in multiple dimensions. Trails and greenways stimulate *economic health* by revitalizing local economies, spurring small business development, increasing property values, and improving quality of life. They contribute to *environmental health*, protecting precious open space and facilitating flood control while encouraging walking and biking as active modes of transportation that reduce air pollution, traffic congestion, and climate change. They contribute to *social health* by providing public space to enjoy the company of family and friends. They contribute to our *personal health* by creating safe, pleasant places to either enjoy active lifestyles or simply quiet the mind and refresh the spirit.

In short, greenway development is a single place-making strategy that produces multiple benefits by improving the economic, social, and environmental health of a place and the personal health of its residents. It is now understood that trails and greenway systems are essential to thriving twenty-first-century communities because they help to create *healthier places for healthier people.*

Despite this new consensus that greenways are imperative, the hard truth is that America lacks an integrated federal-state-local greenway delivery system. When a decision is made to build or expand a highway, a nationwide system is in place to deliver that infrastructure. Nothing comparable exists for greenways. Each project is a "one-off," requiring greenway advocates—both inside and outside government—to navigate an ill-defined process comprising a series of complicated technical, political, and financial steps, such as detailed planning, building public consensus, and securing funding.

This lack of a standard greenway delivery system is exactly why the book you hold in your hands is so vitally important. Chuck has distilled a career's worth of successful greenway development case studies in a single volume with the capacity to both inspire and inform. The lessons learned and detailed in this volume will assist advocates in moving up the greenway development learning curve in order to replicate the successes in this book in communities across America.

For example, how greenways were used in Grand Forks, North Dakota, as the central feature of a flood mitigation effort along the Red River; in Miami, Florida, where, under Chuck's leadership, a neglected landscape has been transformed into a community asset fueling much-needed improvements; in Northwest Arkansas, a regional community that has built a greenway to expand its economy and diversified transportation choice; and in Grand Canyon National Park where a greenway solution improves the visitation experience for millions of people.

But what excites me most about this book is how these individual case studies build toward the compelling vision of the final chapter: A National Greenways System. This book could not be more timely because I am convinced that our growing movement is entering a new phase that will dramatically reshape the American landscape for the next century.

In the last thirty years, America has invested billions of dollars to create thousands of miles of trails and greenways. The vast majority of these greenways are well loved and heavily used. But we have not yet maximized their full capacity to improve the health of America's communities because they were originally conceived as individual units, not as segments of larger systems designed to maximize the numerous benefits produced by greenway systems.

Because the power of connectivity was little understood, these greenways were not designed as networks to maximize use. When connectivity improves in urban or suburban settings, usage soars as more local people can safely and conveniently reach more nearby destinations on foot or bike. When connectivity improves on greenways that link small towns in a scenic rural area, the greenway itself becomes a destination that attracts visitors—and their tourism dollars—from far and wide.

Regardless of the setting—urban, suburban, or rural—connectivity can dramatically increase usage, and increased usage, in turn, is the key factor in producing even more of the numerous economic, social, environmental, and public health benefits that contribute to healthier communities.

With thousands of miles of trails and greenways now on the ground, it is not hard to imagine a future in which these individual pieces are connected to create the National Greenway System envisioned by Chuck in the final chapter. But it is also not hard to imagine the voices of naysayers who will claim that such a vision is simply too expensive. For them, there are two responses.

First, I am convinced that we are at a "tipping point" when relatively small investments to improve the connectivity of existing greenways promise to produce outsized returns because we will reap the full benefit of past investments. We have the opportunity to make smart investments that systematically close gaps, improve access, and tighten connections among existing greenways, thereby creating systems that greatly increase the economic, social, environmental, and public health benefits for relatively small investments.

Second, this opportunity to achieve high rates of return with relatively modest public investment is also consistent with current limits on public infrastructure budgets. At a time of fiscal constraint, investments of a much smaller magnitude than traditional highway projects—but with a much higher rate of return—offer the opportunity to produce a lot more public benefit with far fewer dollars. You could build several regional greenway systems for the price of a single highway interchange. We could create a National Greenway System that would transform America with as little as five cents of every federal surface transportation dollar.

With the cost so low and the return so high, it is imperative that we seize this moment and create something for which future generations will be eternally grateful.

Keith Laughlin
Former president and CEO of Rails-to-Trails Conservancy

THE GREENWAY IMPERATIVE

Introduction

Cyclists ride on a portion of the Grand Forks Greenway in downtown Grand Forks, North Dakota. Source: Greater Grand Forks Convention and Visitors Bureau.

IN THE FALL OF 1997, I was charged with leading part of the disaster recovery efforts in two flood-ravaged communities—Louisville, Kentucky, and Grand Forks, North Dakota. My client in Louisville, the executive director of a public utility that managed the community stormwater system, wanted to use greenways to resolve problems associated with repeated flooding. In Grand Forks, the North Dakota congressional delegation asked the U.S. Army Corps of Engineers to employ a greenway solution as part of the flood recovery process. From my perspective, these disaster recovery projects marked a shift in national perception regarding the function and value of greenways. The greenway was not viewed simply as a linear park, a recreational amenity, or a pathway through the woods as a component of the community park system. The landscape imperative for greenways had changed. Communities were now interested in building greenways to be more resilient to a dynamic and rapidly changing climate, to be more sustainable by conserving rather than exploiting natural resources, and for their economic return on investment. My clients in Louisville and Grand Forks were interested in the ability of greenways to save lives, protect public and private property from flood damage, and resolve critical public health, safety, and welfare issues.

My introduction to greenways began in 1984 when I was employed as the City of Raleigh Greenway Planner. In the mid-1980s, the term "greenway" was not well known and not widely used across the country. There were few communities with constructed greenways and less than a handful with adopted greenway system plans. Raleigh, North Carolina, launched its Capital Area Greenway System in 1969 based on a community plan that defined greenways as the "green fingers" of the city park system. In 1972, North Carolina State University graduate student Bill Flournoy's thesis gave purpose, form, and direction to the greenway plan as multiobjective corridors, established in part to combat urban flooding. More than a decade later, by the time I was hired to run the city program, greenways were viewed by most residents as a subset of the community park and recreation system. The primary purpose of my job was to build urban trails, as defined in the original city greenway plan, to achieve 200 miles of constructed trail by the year 2000. Not long after I began work as the greenway planner, I viewed the opportunity of greenways in a more holistic manner, similar to how Flournoy described them in his thesis. Raleigh's greenways provided pathways for bicycling as alternatives to automobile transportation, conserved natural landscape buffers throughout a community, and protected native species of plants and habitat for animals. With ample vegetation, they helped clean the air of pollutants and trapped sediment before it reached streams and rivers, and when they were as wide as floodplains, helped absorb excess rainwater during storm events. From my perspective, greenways were much more than a path through the woods and addressed more than just the recreation needs of Raleigh citizens.

Today, I continue to find that people across America still think of greenways in constrained terms, again considering them to be, for the most, trails, pathways, or travel corridors for hiking and bicycling. This view of greenways shortchanges their true function and value and is a departure from their historic foundation. So what exactly is a greenway? Why should communities build greenways? How and why do people benefit from greenways?

The classic definition of a greenway is as follows: "Greenways are linear open space established along a riverfront, stream valley, or ridgeline, or overland along a converted railroad corridor, historic canal, scenic road, or other route. Greenways conserve open space, protecting water, vegetation, soil, and natural habitat essential to healthy and sustainable ecosystems. Greenways normally include shared-use trails that serve to link people to community resources, providing health and wellness, recreation, and transportation benefits. Greenways have become an essential landscape of successful, engaged, and progressive communities."[1] This definition is taken from *Greenways for America*, authored by Charles Little. In 1989, Little was hired by Patrick Noonan, founder and chairman of the Conservation Fund based in Washington, DC, who asked Little to chronicle the emerging American greenway movement. Noonan was a member of President Ronald Reagan's 1985 Commission on Americans Outdoors and was convinced that a groundswell of interest and support for greenways was building in rural communities, the suburbs, and large cities across the nation. Little devoted the last chapter of his book to the future of the greenways, concluding with the following statement: "greenways come into being one footstep at a time through individual choice and collective action. And who is to say that such modest steps as these will not begin a historic journey with ethical possibilities unimaginable."[2]

As I reflect on my thirty-five-year professional career devoted to planning, designing, and developing greenways in more than 250 communities located in thirty-six states, I understand and appreciate that our nation has undertaken a historic journey during the past three decades, accomplishing what some thought was unimaginable. Greenways have become a well-known landscape typology for communities large and small. They do in fact help save lives and reduce impact from flooding; they have become essential green infrastructure, while at the same time shaping the way people travel within their communities, supporting active living, conserving thousands of acres of irreplaceable greenspace that otherwise might have been developed, and boosting the economies of cities and towns where they are developed. The common thread and foundation for the vast majority of greenway projects are the people who share a vision for the future and the land that they have chosen to steward. This is what makes the American greenway movement an important chapter in our nation's history.

There are more than a dozen published books devoted to the subject of greenways, most of these written in the past twenty-five years. The majority of these books define and describe the technical, scientific, and design theory for greenways. These books were created to appeal to planners, landscape architects, engineers, academicians, and others who wanted to know more about the technical aspects of conserving resources and building greenways. This book, *The Greenway Imperative*, is an intentional departure from my previously published technical resources and presents the human side of greenway planning and design. To better understand the imagined and unimagined benefits that greenways have bestowed throughout America, this book describes a few of the most interesting greenway projects that I have been blessed to work on, with the goal of sharing the stories of the people, landscapes, and imperatives that made these projects come to life. *The Greenway Imperative* represents my personal reflections on a career that has connected me with extraordinary people who led important greenway initiatives. These stories describe the achievements of an individual, a family, a group of people, and a community. What motivated them to develop greenways? What challenges did they overcome? How have these projects shaped the American landscape in terms of public health, transportation, conservation, historic preservation, and economic development?

I provide the intimate detail about how each project evolved—from vision to reality. What were the different paths taken to greenway development and how were landscapes transformed into economically successful green infrastructure projects through lengthy, often challenging processes? Each chapter reveals the back story behind the subject, ideas, and concepts that led to the initiation and making of a greenway. Within each story, you will learn how these projects were shaped by geopolitical forces, funding, landscape resources, and community needs.

I begin with a story about philanthropy and the gift of a greenway. Chapter 1, "A Close Family Legacy," profiles Anne Springs Close, an award-winning and relatively unknown conservationist who, with her family, during the past thirty years, was able to conserve and open for public use 2,100 acres of land in Fort Mill, South Carolina. The Anne Springs Close Greenway celebrates the natural and cultural resources of southern life. Programmed with events throughout the year, it is one of the best examples of private land conservation in the United States.

During the past three decades catastrophic floods from supersized storm events have impacted communities across the United States, resulting in heightened awareness of the need to plan, design, and build sustainable communities resilient to the impacts of flooding. Grand Forks, North Dakota, and East Grand Forks, Minnesota, were ravaged by a 500-year flood event in May 1997. Having rejected a U.S. Army Corps of Engineers proposal to build a river bypass channel, the communities pursued a greenway solution to flood control. Twenty years later the 2,200-acre Red River Greenway has become the centerpiece of life in the community, which has made the community resilient to flooding, and a significant factor in attracting new business and industry to these out-of-the-way upper Midwest communities. Chapter 2, "Come Hell and High Water," takes you through the recovery process, master plan, and building of the Red River Greenway.

Our planet is drowning in postconsumer and postindustrial trash. Earth's oceans are the dumping grounds for millions of tons of plastic waste. Solid waste landfills are overflowing. Rivers and streams are polluted with discarded refuse. What steps can all of us take to reduce, recycle, and reuse the trash we generate every day? Chapter 3, "Turning Trash into Trails," describes efforts of one community during the early 1990s that set out to prove that common household and postindustrial trash could be resources for building greenway trails, bridges, benches, and educational signs. Described are the myriad of products used to build the Swift Creek Recycled Greenway in Cary, North Carolina. How have these products performed through the years? What are the current barriers to the use of products made from recycled trash?

In 1997, overcrowding at the Grand Canyon was a public safety concern as visitors were being squeezed into undersized overlooks and fights were breaking out over limited parking spaces. The National Park Service needed a new approach for how people visit and experience this UNESCO World Heritage landscape. For more than a decade (1997–2007), I worked with the National Park Service and a philanthropic organization to develop rim-top greenway trails on the North and South Rims of Grand Canyon National Park. In addition to relieving overcrowding, these greenways now provide alternatives to automobile travel throughout the park and make rim trails accessible to all people, regardless of their ability. Today, the network of greenway trails has positively changed the way that tourists experience the park and has become a model for what other national parks can accomplish. Chapter 4, "Something Grand," tells the story about the people and events that influenced the design and development of these rim trails.

The conservation of land and water is an urgent imperative in America's fastest-growing communities, even at the edge of the Mojave Desert in southern Nevada. Las Vegas is one of the fastest-growing cities in the world, achieving a population of 2.2 million in less than 100 years. In 2004, city residents completed an open space and trails plan to protect portions of the native desert ecosystem. This plan launched other efforts that in turn gave rise to a land conservation initiative across southern Nevada. The resulting network of greenways, parks, and open space defines a new legacy of resource protection amid the uninhibited consumption that defines Las Vegas. The saying "What happens in Vegas stays in Vegas" is one of the world's most successful marketing slogans. Chapter 5, "Open Space in Vegas—It's a Sure Bet," tells a story worth sharing with the entire world.

Sometimes the greatest asset within a community is the undervalued landscape in full view of everyone. The City of Miami is named for a river that flows through the heart of its downtown. However, few residents of Miami realize that the river exists, much less that their community is named for the river. That all began to change in the late 1990s when the Trust for Public Land began working with the Miami River Commission to conserve and restore this iconic natural and cultural resource. Chapter 6, "Miami Means 'Sweet Water,'" describes the obstacles that were overcome in crafting a greenway plan for the river, including resolving lack of access to the riverfront, transforming neglected adjacent riverfront properties, reducing crime and illegal drug trade, and cleaning the river of sediment and pollution.

In 2004, the voters of Charleston County took a "leap of faith" to approve a ½-penny sales tax to fund countywide transportation improvements and greenspace conservation. Voters passed this ballot initiative without the benefit of a master plan for how to spend the money. In 2005, using little more than the one-sentence description from the ballot initiative, I worked with a citizens group to craft the plan and strategy for setting aside land and water that would satisfy the spirit and intent of the citizen vote. Chapter 7, "Lowcountry Life," describes the process that Charleston County implemented to launch one of America's most successful land conservation initiatives—protecting 40,000 acres of land in little more than a decade within one of the nation's fastest-growing communities.

To ensure its community residents are healthy and productive, and that the cities and towns of northwest Arkansas are economically competitive on a global scale, the Walton Family Foundation, based in Bentonville, Arkansas, decided to invest in greenway development. The foundation has a progressive vision for their Midwest, historically agricultural communities, to transform them into sought-after destinations for mountain biking and outdoor activities. The centerpiece of this vision is the Razorback Greenway, a thirty-six-mile urban greenway that links together the cities and towns of the region. Chapter 8, "Callin' the Hogs," describes the process for how this greenway was envisioned, funded, planned, designed, and constructed, all within a compressed timeframe.

Rural economic development is challenging no matter where it occurs—whether in the American heartland or other nations around the world. The nation of Belarus is a significant trading partner of Russia, supplying important agricultural products. The government of Belarus, however, must diversify its agricultural economy to offset unexpected downturns. In August 2012, Country Escape, a nongovernmental organization, engaged me to work with the citizens and government of Belarus on an agrotourism strategy. During my ten-day visit, funded by the U.S. Embassy in Minsk, I met with government, nongovernment, and business leaders to define ways greenways could support economic development and tourism. Chapter 9, "White Russia," summarizes the resultant greenway-oriented agrotourism strategies that evolved and which of these are transferable in rural American communities.

Why do we care that long-distance trails are planned and developed? Why should America invest in a national network of greenways and trails? Chapter 10, "America's Longest Urban Greenway," profiles the work of a small group of people, who in the early 1990s gathered in New York

City to envision one of America's longest greenways. Over time the vision for this trail extended 3,000 miles from the Maine-Canadian border to Key West, Florida. Today, the East Coast Greenway provides nearly 100 million people across fifteen states and 450 cities with a safe, multimodal network of trails that comprises the nation's longest continuous urban biking and hiking route. Since its founding, this greenway has connected people to native and cultural landscapes, stimulated economic development in big cities and small towns, created resilient public infrastructure, and encouraged equitable, inclusive, and innovated trail development that is free for the public to use. This is the story of the East Coast Greenway as the urban equivalent of the Appalachian Trail. It may best be understood as the "human powered Interstate 95."

For the past 100 years, America has been slowly building a national network of interconnected trails and greenways. We have reached a tipping point in which completing that system is within the realm of possibility. In 1919, landscape architect Warren Manning was one of the first to envision and record a National Plan for an interconnected network of trails and greenways across America. Significant federal legislation in the 1960s and federal funding in the 1990s established a foundation that helped launch the modern American greenway movement. From 1985 to 2019, our nation has made significant progress in realizing Manning's vision, adding more than 50,000 miles of greenways to the American landscape. Chapter 11 explores the vision, framework, and steps necessary to complete the bold undertaking for a national greenway system.

How might an interconnected network of greenways address the consequences of a warming planet? Responding to this life-altering reality, we need to adjust our thinking to the notion of greenways as "gene-ways"; critical migratory routes for plants and animals through interconnected north-south and east-west corridors of land dedicated to this purpose. As "a nation of trails," American history is filled with important routes of travel defined by significant events. Cultural trails, an emerging component of America's national trails network, provide the opportunity to connect with our unique and at times painful history and connect us with landscapes of celebration and achievement. In recognition of our increasingly diverse culture, the work associated with greenway planning, design, and development needs to include the input of all people, regardless of ethnicity or heritage. This work must address equity, environmental justice, and the celebration of culture as beneficial outcomes of greenway development.

Charles Little concluded that "to build a greenway is to build a community." We are a very different nation of communities from the one that I was born into in 1959, the one in which I grew up as a child, and the one I knew when I began my practice in the mid-1980s. Looking to the future, the planning, design, and building of sustainable, resilient communities will be among the most significant challenges we face as a nation. Communities we build for the future must adapt to the realities of a dynamic ever-changing environment. They must also satisfy human need for connections to nature.

Fundamentally, greenways are about connections—to the land, to our communities, to each other. Greenways can help us to achieve resilient communities that respond to public health, safety, and welfare concerns, become important elements of a national interconnected framework of green infrastructure, and serve as keystone landscapes in all fifty states. I hope by reading *The Greenway Imperative*, you are inspired from the lessons learned to take action in your community to conserve natural and cultural resources, build safe, resilient, diverse and sustainable communities, and make changes in lifestyle and community development patterns that enable all of us to live healthfully in greater harmony with Earth's bountiful resources.

1

A Close Family Legacy

Anne Springs Close Greenway, Fort Mill, South Carolina

A family hikes along a portion of the Anne Springs Close Greenway in Fort Mill, South Carolina. Source: Leroy Springs and Company.

WE SAT HIGH ON HORSEBACK that cold and blustery winter morning, looking over the remains of what was once a majestic old growth forest. The air had an odd smell, a pungent mixture of sweet oak sap, pine pitch, and musky South Carolina red clay. It was an unnatural smell for a forest illuminated by an extraordinary amount of light—as there was no canopy of branches to filter the sun. Naked wood trunks framed the views depleted of their mature crown, which had been snapped by the 100-mph winds of a Category 4 hurricane. The pain of her loss was etched deeply in Anne Springs Close's weathered face and resonated in her carefully chosen words. "I have hiked through this forest all my life and have never seen such devastation." Indeed, the forest looked as though someone had grabbed several handfuls of matchsticks and indiscriminately tossed them across an open field. Eighty-foot oak trees lay crushed and splintered on the forest floor, one atop another. It was a battlefield, not a forest, and we were surveying the remnants of nature's fury. Hurricanes are capable of such amazing destruction and devastation. Hurricane Hugo was one of the most destructive in decades, and it had wrecked a forest that meant everything to Anne Springs Close. These woods held a lifetime of memories and childhood experiences, walks with her husband, children, and close friends, horseback rides across Steele Creek, and summer morning swims in the millpond. All those memories seemed not just distant but swept away with the gale force winds that toppled the forest.

Anne Springs Close is a celebrated conservationist and has always been a proactive person. She fully understood that the devastation of Hugo would never be the lasting legacy of this forest and her memories of this special landscape. She had vision and determination to mend this land, restore it, and share its bounty with others as she had always known it to be: a place of solace, beauty, inspiration, adventure, and learning. With this vision and determination Ms. Close and her family established one of the America's most significant, privately managed greenways, a legacy landscape known today as the Anne Springs Close Greenway.

A Meeting by Chance

I met Ms. Close for the first time at the National Recreation and Park Association Conference in Baltimore, Maryland, in October 1991. I was delivering a presentation at the conference, titled "Green Ideas: A Road Map to Greenway Planning." Ms. Close was attending the conference on a mission—to find an experienced greenway designer to help her complete a master plan for a 2,100-acre greenway on property owned by her family. Years later, Ms. Close confirmed with me that our meeting in Baltimore might have never happened. She and Bob Reid, president of Leroy Springs and Company, scanned the conference program guide when they arrived. There were several presentations that week on the topic of greenways. She contemplated the presentation descriptions and speakers and selected mine as the one to attend. To this day, she is not sure why she picked my presentation over the others. However, she firmly believes that it was ordained that we meet that afternoon in Baltimore, that other forces were in fact at work, and that God had always intended that we would work together to craft a master plan for her greenway.

Ms. Close and Mr. Reid approached me after my talk and introduced themselves. I immediately knew who Ms. Close was, but I waited patiently for her to explain the project that she had in mind and her reasons for attending my presentation. What Ms. Close did not know, at the time, was that there was a family connection between the two of us. My mother's uncle, Murphy Gregg, was her father's (Colonel Elliott Springs) right-hand man. My great-aunt Olive Gregg, Murphy's wife, hosted a November 1946 wedding party for Ms. Close and her husband, Bill Close, at Springfield, one of the historic homes on the Springs-Close family property in Fort Mill, South Carolina. It was Olive Gregg who worked with Colonel Springs to restore and modernize Springfield. My family and I visited the Greggs in the summer of 1963 when I was four years old. During my years at North Carolina State University, the Greggs became my surrogate family, where I spent many of my holidays and weekends—a stone's throw from historic Springfield. I knew of this history, and it was all that I could do to calmly explain to Ms. Close and Bob that not only did I know who they were, but that I was already somewhat familiar with the property and the potential for a greenway. When I finished my explanation, Ms. Close looked me in the eye, wobbled a bit, grabbed a chair, and proclaimed, "I really need to sit down!" She was momentarily shaken but quickly composed herself and responded in a firm tone, "Well, when can you come to Fort Mill to look at the property and begin work on a greenway plan?"

That fortuitous meeting in Baltimore was as much a gift to me as it was to Ms. Close. It has been my honor and privilege to work with Ms. Close, her family, and Leroy Springs and Company for the past thirty years to assist in the planning and design of the Anne Springs Close Greenway. I am often asked: "What is your favorite greenway project?" and that is a nearly impossible question to answer. This book is filled with stories of some of my favorites. However, it is hard to deny the spiritual connection that I have come to know through my involvement with the Anne Springs Close Greenway.

Working with Ms. Close to plan and design her greenway affirmed for me a spiritual connection to the land, to a special place deep in my soul that no other professional project has touched. So maybe, just maybe, Ms. Close is correct. God's plan *was* for us to meet in Baltimore, and the two of us *were* always destined to work together to craft a plan of stewardship for one of America's special landscapes; to preserve in perpetuity ecosystems and irreplaceable cultural assets for future generations to interact with, learn from, and cherish.

A Southern Mill Town

Fort Mill, South Carolina, is not where one might expect to find one of America's most significant privately owned and operated greenways. Fort Mill is a small southern community, a thirty-minute drive south of downtown Charlotte. Situated in the midst of a rapidly developing area of South Carolina, Fort Mill is a quintessential American mill town in size, character, and function. The origins of Fort Mill date to colonial America, when Webb's Mill (a grist mill) was established on the banks of Steele Creek. Scotch-Irish immigrants established a small outpost that was located between a British fort designed (but never built) to protect the Catawba Indian Nation and the grist mill, hence the name Fort Mill.[1] The settlement was home to John Springs (1782–1833) and Samuel Elliott White (1803–1865). The union of the families produced Elliott White Springs, who transformed the Fort Mill Manufacturing Company, founded by his grandfather, into Springs Industries, one of America's most successful cotton manufacturing enterprises.

The greenway is the product of three contributing factors: a timeless parcel of land that harbors an intact southeastern ecosystem, the Springs-Close family who have been stewards of this land for more than 200 years, and a unique cultural heritage that reflects the founding, evolution, and development of the United States.

A Family of Achievement

Anne Springs Close will be the first one to tell you that she has lived a privileged life. She has used her gift of privilege to forge a life dedicated to conservation, community betterment, and philanthropy. These values were instilled in her by her parents, Colonel Elliott White Springs and Frances Ley Springs. Colonel Springs was a decorated World War I double ace pilot. He authored *Warbirds: The Diary of an Unknown Aviator* (1927),[2] regarded among the most important writings of World War I aviation history. Upon the death of his father in 1931, in the darkest days of the Great Depression, Elliott Springs took control of his father's failing textile mills, the Springs Cotton Mills, and during the course of a distinguished professional career that lasted to his death in 1959, transformed the family business into one of the nation's most successful industrial enterprises.

My great-uncle, Murphy Gregg, was tasked with implementing Colonel Springs's vision for Springs Industries and the sprawling 7,000-acre Springs family estate. Murphy came on board to work closely with the Colonel and Springs Farm manager R. F. Palmer, to implement a land stewardship program for the estate. These are facts I have only come to appreciate many years later, as Murphy never divulged to me his relationship with the Colonel, nor the work that they accomplished together. It was Ms. Close's son, Elliott Close, who confirmed that Murphy Gregg, by virtue of the faith and trust vested in him by Colonel Springs, was one of the most respected and influential men in the "up-country" area of South Carolina.

Elliott White Springs built an industrial empire that included manufacturing, banking, cotton, oil, rail transportation, and insurance. The Colonel is, to this day, remembered for his humor and zest for life, best reflected in a 1949 Madison Avenue advertisement for Springmaid sheets and linens. The ad featured an exhausted Native American male lying in a hammock made from Springmaid sheets with an attractive young maiden climbing from the hammock. The tag line below the graphic, "A buck well spent on a Springmaid sheet," is to this day one of the more ingenious, notorious, and revolutionary commercial advertisements ever crafted. The ad made Springmaid sheets and linens a household name. The Colonel's instinct for business and his ability to envision the future fueled the growth of his industrial empire. Colonel Springs overcame many challenges as an aviator, writer, businessman, husband, and father. When he died suddenly in 1959, his legacy of achievement was passed on to his son-in-law, Bill Close, and his daughter, Anne.

One of the enduring qualities of Colonel Springs was his benevolence and charity. It is recorded that from 1925 to 1959, the Colonel bequeathed more than $15 million of his personal wealth to support health care and college scholarships for the people of Lancaster County, Chester County, and the community of Fort Mill. Elliott White Springs established several important philanthropic organizations, including Leroy Springs and Company (LSC), begun in 1938 and named for his father, and the Springs Foundation, Inc., begun in 1942. The LSC operated various holdings of the Springs Close family, including Springmaid Beach and Springmaid Mountain, and today the Anne Springs Close Greenway. As of 2017, the LSC awarded more than 5,000 scholarships and loaned more than $2.6 million. The Springs Foundation, now known as the Springs Close Foundation, has awarded more than $100 million to a variety of community and educational causes to improve the quality of life for residents of Chester, Lancaster, and York counties.

The Colonel cared deeply for the people who worked in his mills and he went to extraordinary lengths to ensure that they had access to amenities that would improve their quality of life. He built recreation centers in each of the mill towns where Springs Industries operated. He owned and operated, for the benefit of mill workers, Springmaid Beach (opened in 1949 and sold in 2014), an affordable oceanfront resort in Myrtle Beach, South Carolina. Colonel Springs was the epitome of the benevolent American industrialist. He instilled these values in his daughter, Anne, and his son, Leroy "Sonny" Springs, who died at the age of twenty-two tragically in an airplane crash on Mother's Day in 1946. When it came time to formally establish the Anne Springs Close Greenway as a gift to the residents of Fort Mill, surrounding counties, and the State of South Carolina, Ms. Close and her family used the resources of Leroy Springs and Company to facilitate the transaction. The Colonel's community vision, interest in land resource stewardship, and lifetime of philanthropy had come full circle and is realized by the extraordinary and everlasting gift of the greenway.

Connection to the Land

Anne Kingsley Springs Close was born in 1925 and raised in Fort Mill. As Her German au pair, Toni Dehler (Miss Toni) raised Anne in accordance with European values that fostered an appreciation of native ecosystems and included daily hikes through the countryside. This upbringing had a profound influence on Ms. Close's life and her conservation,

Anne Springs Close is the matriarch of a conservation-minded family and namesake of the Anne Springs Close Greenway. Source: Leroy Springs and Company.

recreation, and community wellness pursuits. Ms. Close vividly recalls walks with her brother, Sonny, and Miss Toni that would at times stretch to more than ten miles and involved circumnavigating the vast estate being assembled by her father. Ms. Close attended Fort Mill public schools as a child, Ashley Hall in Charleston and Chatham Hall in Virginia, and Smith College in Massachusetts. She returned to Fort Mill, South Carolina, and married Hugh William "Bill" Close in the fall of 1946.

Anne Springs Close is an extraordinary person who revels in the spirit and adventure of life each day. She is a hands-on person who has been involved in notable South Carolina park and conservation projects, including the Palmetto Trail, which she envisioned as a hiking trail extending 400 miles from the Upstate to the Lowcountry of South Carolina.

She has been instrumental in the success of the Nation Ford Land Trust and York County Forever, both of which are dedicated to the conservation of environmentally sensitive lands. She has also been the director of summer equestrian camps, which conduct weekly riding lessons for disabled children at the stables near her home. She has served as chair of the National Recreation and Park Association Board of Trustees, director of the American Farmland Trust, and trustee with the Wilderness Society. Perhaps most important, Ms. Close has carried on the tradition of benevolence established by her father in supporting the operation of the greenway during the past two decades from her personal endowment. Ms. Close is the recipient of the 2001 Cornelius Amory Pugsley Local Medal Award for her "outstanding contributions to the promotion and development of public parks in the United States." Other Pugsley award recipients include Stephen Mather, the first director of the National Park Service, Lady Bird Johnson, Stewart Udall, former Secretary of the Interior, and Laurance S. Rockefeller, just to name a few of the more than fifty recipients to date.

Ms. Close's husband, Hugh William "Bill" Close, rightfully deserves tremendous credit for his work building upon Colonel Springs's vision, leadership, and accomplishments, and growing Springs Industries into a multimillion-dollar global company. Bill Close continued the philanthropic endeavors of his father-in-law until his untimely death in 1983, serving simultaneously as president and chairman of the Springs Close Foundation, Leroy Springs and Company, and the Frances Ley Springs Foundation.

Bill and Anne raised eight children, all of whom have achieved remarkable success. Having worked with all her daughters and sons to plan and design the greenway, I know that each of them individually and all of them collectively embody the spirit of adventure, compassion for humanity, and dedication to stewardship that was instilled in them by their parents and grandparents. Crandall Close Bowles, Frances Close Hart, Deacon Leroy Springs "Buck" Close, Patricia Grace Close, Elliott Springs Close, Hugh William "Will" Close, Derick Close, and Katherine Anne Close, M.D., collectively believe that the Anne Springs Close Greenway is one of the single greatest achievements of the Springs Close family. The 7,000-acre Springs estate was willed by Colonel Springs to the eight children and not to the Colonel's daughter. Ms. Close had to seek both permission and agreement among all eight children that conserving this land and establishing a greenway was in the family's best interest. Ms. Close's children agreed, and without their support, there would be no greenway. The children named the greenway in honor of their mother, and formally declared their dedication to the long-term stewardship of the land and cultural assets by conveying a conservation easement across the entire greenway in August 2007 "to preserve and protect the cultural, ecological, educational, and where applicable, the agricultural, natural open space and scenic features of the property and to prevent any use of the property that will significantly impair or interfere with the conservation values of the property."

A Legacy Landscape

The land that comprises the 2,100-acre Anne Springs Close Greenway is the combined estates of John Springs III (Anne Springs Close's 2nd great grandfather) and William Elliott White (Anne Springs Close's 2nd great grandfather). The entirety of the greenway is situated within a portion of the fifteen-square-mile, 144,000-acre reservation that was established for the Catawba Indian Nation by the South Carolina legislature in 1763 (Treaty of Augusta). All this land was transferred to white settlers by the State of South Carolina in 1840 (Treaty of Nation Ford) in exchange for promises of cash and new reservation land for the Catawba in North Carolina. Messrs. Springs and White, who were tenants and paid rent to the Catawba, were two of the many beneficiaries of the land transfer.

The land within the greenway has been under active stewardship, cultivation, and production for more than 300 years. English Revolutionary War General Lord Cornwallis described this area of South Carolina as parklike "with no underbrush but greensward as far as the eye can reach." The Catawba Indians carefully managed forest undergrowth to encourage grasses and other vegetation that would provide food for bison and deer. This land stewardship continued under the Springs and White families, who farmed and managed the forests for agricultural production. Much of the forest toppled by Hurricane Hugo in 1989 had, for the most part, been undisturbed by human intervention. Other than the gristmill and associated millpond adjacent to Steele Creek, the creek and the surrounding floodplain had also generally remained undisturbed for decades.

Greenway Vision and Master Plan

The Anne Springs Close Greenway does not fit the classic description of a "greenway" as defined by Charles E. Little in *Greenways for America*. The greenway, roughly 3 miles wide and 4 miles long, is an irregular, rectangular-shaped environmental and cultural park rather than the long, skinny, linear park associated with other American greenways. At 2,100 acres in size, the greenway is three times the size of Central Park in New York City.

Ms. Close's lifelong desire and interest in establishing the greenway was first depicted in a 1987 land use plan for the 6,700-acre Close family landholdings prepared by The Conservation Fund and D. R. Horne and Associates. LandDesign, Inc., a planning and landscape architecture firm based in Charlotte, North Carolina, was subsequently retained by the Close family to prepare a master plan for the land slated for future residential and commercial development. The Clear Springs Master Plan, prepared by LandDesign, led to the development of the residential and mixed-use communities of Baxter and Springfield, adjacent to the greenway.

Beginning in 1991, I worked with Ms. Close and Leroy Springs and Company to define a 100-year vision, 20-year master plan, and 5-year action plan for the greenway. The master plan had to properly account for 180 years of family ownership and resource management. It also had to define a strategy of introducing public access to lands and waters that had heretofore supported limited use and accommodate the development of new facilities to support access, use, and environmental education. A balance between use and conservation was of paramount importance and concern and had to be appropriately resolved by the master plan.

The master planning process was based on renowned Scottish landscape architect Ian McHarg's "Design with Nature," which emphasizes the importance of ecological planning and sustainable design. McHarg's concepts are the basis for modern Geographic Information Systems analysis and mapping that is employed throughout the world today. Using

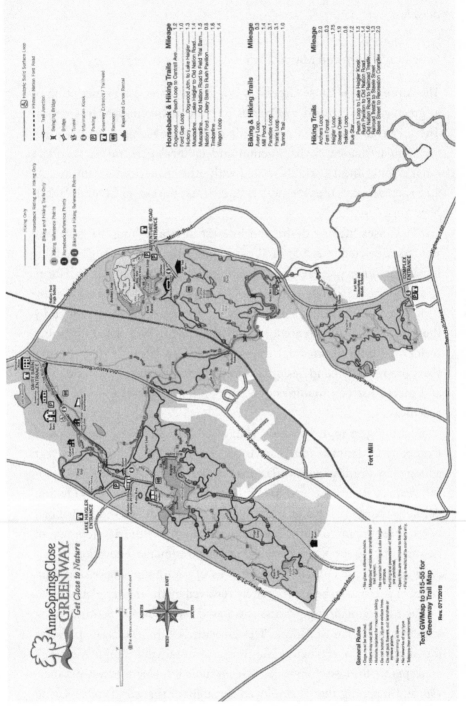

Anne Springs Close Greenway trail system map. Source: Leroy Springs and Company.

McHarg's process, I mapped out the topographic, hydrologic, and current land use components of the 2,100-acre project study area as individual layers of crucial ecological information. The cultural and historic components of the property were crucial to the design program, and my team mapped and described each of the most important architectural elements: the historic homes, farm buildings, and other heritage structures. Then by overlaying and combining each separate layer, an opportunities and constraints map was defined for the proposed greenway, which served to inform the final design program. The Final Greenway Master Plan was based largely on the ecological and cultural characteristics of the land.

It was important to establish long-term objectives for the greenway as a protected and conserved landscape. The 100-year vision reflects the Close family's values and desires for the greenway, as originally articulated in 1990 by Ms. Close: "The Greenway will provide a beautiful parklike area for a wide variety of low-impact recreation activities, while maintaining a natural habitat for wildlife. By interpreting and protecting the Steele Creek watershed and adjacent natural resources, the Greenway will serve as a buffer against urban encroachment, protecting the community for generations to come."

Leroy Springs and Company was charged with operating and managing the greenway. The master plan formally adopted in October 1992 was designed to provide the family and LSC with the tools and program of action to balance improved public access and use with resource conservation, protection, and stewardship.

The twenty-year master plan called for using the existing farm manager's house as the Greenway Information Center, which would house daily operations and management. More than 36 miles of trails were proposed for development throughout the greenway, with 14.6 miles dedicated to equestrian use, 5.5 miles for multipurpose trails, 7.4 miles of hiking trails and 9.4 miles of sidewalk and bicycle trails.

Lake Haigler, the largest body of water on the estate, named for the Catawba Indian chief Nop-ke-Hee (known in English as King Haigler), supports shoreline fishing, canoeing, and kayaking as well as primitive campsites around the perimeter of the lake. Campsites are furnished with fire rings, picnic tables, and a level pad for a single tent. Formal entrances to the greenway were strategically situated to accommodate public entry.

To address restoration of the old growth forest smashed by Hurricane Hugo, I enlisted the help of Dr. Robert Bruck, a national and international expert in forestry management from North Carolina State University to advise the Close family and LSC on regenerating the forest system. Dr. Bruck produced a forest management plan that served the immediate and long-term needs of the estate.

The 1992 Greenway Master Plan was simple, straightforward, easy to follow, and respectful of what the land had to offer. Much of what has been constructed during the past thirty years came from resources indigenous to the land. To the credit of Ms. Close, her family, and LSC, virtually every one of the recommendations within the twenty-year plan was implemented as originally defined. The only exception was the conversion of the historic Dairy Barn into a proposed Greenway Center as an environmental education center. The Close family and LSC decided to transform the largest building on the estate into an events center, and as a result it has become one of the most popular and sought-after event centers in York County, South Carolina.

Runners participate in a race along the Anne Springs Close Greenway trail system. Source: Leroy Springs and Company.

In 2009, I was retained by Tim Patterson, president of LSC, and the Close family to update the Greenway Master Plan and define a new twenty-year strategic plan. It was gratifying to see how much had been accomplished in the prior twenty-seven years. The greenway had become the beloved resource originally envisioned in the early 1990s. The forest system had completely regenerated, activity centers were fully utilized, and the greenway was a living laboratory of conserved land for new generations of conservation-minded adults and children to explore and learn from.

The strategic plan completed at the close of 2010 envisioned building two needed facilities: a welcome center (opened in 2019) and an outdoor performing arts theatre (opened in 2015). Both these recommendations fit the vernacular and programmatic needs of the greenway. With the addition of these new facilities, the greenway will capitalize on available resources, expand its program offerings, and engage visitors in additional opportunities for education and outreach.

Snapshots of American History

To traverse the landscape of the Anne Springs Close Greenway is to travel back in time through American history. The greenway has preserved, protected, conserved, and restored this timeless American landscape. Five places and structures are worthy of further description: Nation Ford Road, Springfield, White Homestead, the Faires-Coltharp and Graham Cabins, and the Webb Gristmill.

Nation Ford Road

Nation Ford Road is an ancient roadbed established by several Indian nations and originally known as the Trading Path. Nation Ford Road comprises a portion of the historic Philadelphia Wagon Road, north-south highway of Colonial America. The road was used by white settlers as early as 1650 for travel between the James River in Virginia and the Catawba River in South Carolina. An intact remnant of the road has been protected in perpetuity within the greenway and is part of the trail system. When you hike along Nation Ford Road, you are traveling the same ground as Catawba Indian chiefs, hiking in the footsteps of British forces, led by General Cornwallis, as they retreated from defeat at the hands of American patriots, mirroring troop movements of Confederate and Federal soldiers during the Civil War, and retracing the movement of immigrant Americans who were in search of a better life for their families. It is one of the most powerful landscapes to experience firsthand, and one of the last remaining intact sections of the road in its natural context. The greenway segment of Nation Ford Road was listed in March 2007 with the National Register of Historic Places.

Springfield

Situated in the northeast corner of the greenway, Springfield was the home of John Springs III and was built sometime prior to 1806. Springfield is one of the few antebellum homes remaining in Fort Mill, South Carolina, and is one of the oldest structures in eastern York County. It is currently the home of Leroy Springs and Company. My great aunt Olive Gregg worked closely with Colonel Springs to meticulously restore the home and add modern plumbing, central heat, and electrical fixtures.

Springfield is a two-story wood frame structure. The house consists of a gable-roofed block with pedimented gable ends. The exterior of the house is weatherboard painted white, with black trim shutters. It is a striking home, set among native oaks and accented with a boxwood hedge, holly, and magnolia. The interior of the home is breathtaking. The central hall is accented by a double front door surrounded by thirteen panes of stained glass commemorating the original colonial states. The central hall opens to identical parlors on either side. The south wing of the house contains a very large dining room and the north wing features a large bedroom. A new kitchen was added to the home in 1946. The upstairs of the home consists of a central hallway and a series of bedrooms and bathrooms.

Springfield is notable for many reasons. The house served as host for Confederate President Jefferson Davis and his cabinet in April 1865. Davis and five members of his cabinet were fleeing Federal soldiers, with the treasury in tow. They spent several nights at Springfield and conducted one of their final meetings at the White Homestead in Fort Mill. Leroy Springs and Company undertook renovations to the home in the early 1980s to convert it to office space, and in 2010 to restore the front porch to

The historic Springfield house on the Anne Springs Close Greenway in Fort Mill, South Carolina. Photo by author.

its original design and upgrade historic elements that were damaged or in disrepair. The home is listed on the National Register of Historic Places.

White Homestead

While technically not part of the Anne Springs Close Greenway, the White Homestead nevertheless deserves mention as the homeplace of William Elliott White and the childhood home of Anne Springs Close. For those interested in Southern American history, in April 1865, less than two weeks prior to his capture by Federal troops, Jefferson Davis held a final Confederate States of America cabinet meeting on the front lawn of the homestead. Today, the home is privately owned, primarily used for family functions, and not open to the public. The home is Georgian architecture and resembles a manor from Yorkshire, England. The main part of the house was constructed in 1831 and is four stories in height, counting the basement and attic. It is the oldest surviving brick structure in York County. Elliott White Springs and his wife Frances Ley took residence in the home after their wedding in 1922. Colonel Springs added a wing to the home in 1936, and a combination greenhouse and swimming pool in 1953. When Elliott White Springs died in 1959, he willed the home to his eight grandchildren. In 1991, the home underwent a full restoration. It is listed on the National Register of Historic Places.

Historic Log Cabins

There are two historic log cabins within the greenway: the Faires-Coltharp Cabin, circa 1800, and the Graham Cabin, circa 1780. These cabins provide visitors with a better understanding of life in South Carolina during the nineteenth century and contain an outstanding collection of furniture, clothing, artifacts, and antique quilts of the period. The Graham Cabin was the home of William C. Graham, the Reverend Billy Graham's grandfather. The cabin was constructed from hand hewn American chestnut trees and was originally located west of Fort Mill until 1999 when it was relocated to the greenway. The two-story, 240-year-old cabin is in excellent condition and consists of one large room on the ground floor with a stone fireplace, a front porch, and a larger porch that wraps around two sides of the cabin. Other than its historical significance and ties to the Graham family, the cabin is also famous for its link to the plight of the American chestnut. During a period between 1904 and 1959, more than 4 billion American chestnut trees died throughout Appalachia due to a fungal disease. Currently, tours of the cabins are by invitation only and during greenway events, such as Fort Mill Frontier Days in March or Earth Day on the greenway in April. The Faires-Coltharp cabin was constructed around 1778 by Revolutionary War veteran Alexander Faires. The architecture and interior of the cabin depict how eastern American families lived during the eighteenth and nineteenth centuries. At one time, twelve family members occupied the cramped quarters of the four-room cabin. In addition to serving as a historic structure for interpretation, the cabin hosts annual Faires family reunions.

Grist Mill and Millpond

During her childhood, Anne Springs Close used to swim in the millpond adjacent to Steele Creek. That thirteen-acre millpond was part of what is known as Webb's Mill, a grist mill established in 1760 to process corn and wheat, and to saw lumber. John Springs III acquired the mill in 1813 and continued to operate it until 1837. Remains of the grist mill are still visible in Steele Creek, as is the mill race that powered the grinding wheel. In 2009, Leroy Springs and Company constructed a new building, along with a functional water wheel, as a representation of what the mill might have looked like. Inside the mill house are educational exhibits that depict the location and approximate size of the millpond and information that describes how corn and wheat were processed. The mill house is open by invitation and on greenway event days.

Anne Springs Close Greenway Today

The Anne Springs Close Greenway has become a beloved, popular landscape, programmed with activity almost every day of each month throughout the year. In 2017, the greenway was the tenth most visited landscape in the Charlotte metropolitan region, with annual visitation exceeding 240,000 users.

The greenway is home to a diverse array of programs and events, including an annual Earth Day celebration, Bluegrass Festival, Fall Festival, Summer Concert series, winter crafts fair, and Green Gala. The greenway supports an equestrian center where horses can be pasture boarded, competition riders can participate in spring and fall horse shows, and children with disabilities participate in guided horseback rides. Sponsored hikes and formal races are held on the forty miles of natural surface greenway trails for hikers, trail runners, and mountain bikers. Users can participate in summer yoga classes, tour historic cabins, or canoe and kayak on Lake Haigler. Overnight camping is available for individuals or small groups. Visitors are required to pay a daily fee to enter the greenway or can purchase an annual membership. A Greenway Assistance Program (GAP) provides affordable memberships and access to community residents with financial limitations. The GAP is applicable to annual memberships, summer camps, equestrian programs, and school programs.

The greenway partners with the Fort Mill School District to host a variety of environmental educational programs, including summer camps, school field trips such as "science in a backpack," a South Carolina Master

Naturalist program, and area scout troops. In 2017, 16,000 children benefited from the variety of environmental education offerings and more than 4,500 children participated in the summer camp programs. The greenway is supported by a variety of volunteers and members, including Friends of the Greenway, the Ladies Guild, the Colonel's Club, and corporate partners. In 2017, volunteers donated more than 5,600 hours in support of stewardship activities within the greenway.

The Power of One

Anne Springs Close says with great modesty, "I originally set out to preserve a few trees and conserve a landscape that would serve the needs of the community." Thirty years later, the Anne Springs Close Greenway has achieved so much more. This is testimony to the power of a passionate, forward-thinking individual and the impact one person can have on the land, a community, and a region. What has made the greenway successful has been the commitment of the Close family to a holistic and lasting vision for the property, the bold conservation initiative implemented by the family, and the ability of the greenway to become one of the most valued assets in the up-country region of South Carolina. Setting aside some of the most valuable, most marketable, and highly developable land was a bold initiative by the Close family. As a family, they agreed to forgo significant personal economic gain for the sake of conservation. Additionally, in order to operate the greenway in perpetuity they have established a multimillion-dollar endowment that provides an annual income stream to offset greenway maintenance costs. This is a remarkable model of philanthropy that we need more of in the United States.

Importance of Philanthropy

Philanthropy means "the love of humanity." For the Close family, the word has significant meaning and was fully expressed by Colonel Elliott White Springs in his concern and provisions for mill workers. It is the underlying motivation behind the conservation of 2,100 acres of the Springs-Close estate as a greenway, which now defines the family legacy. As a lasting statement of more than 180 years of influence on the community, the Close family wanted to give back and pay forward through the gift of a greenway that celebrates the land and history of the region, conserves precious natural resources, and creates an outdoor venue that enriches the lives of generations to come.

Some of the most high-profile greenway projects in America have been financially supported by philanthropic individuals and organizations. The High Line in New York City, Northwest Arkansas Razorback Regional Greenway in Bentonville and Fayetteville, Arkansas, the Wolf River Greenway in Memphis, Tennessee, and the Bayou Greenways in Houston, Texas, are just a few examples of greenway projects that have received substantial funding from philanthropy.

In October 2018, the Ralph Wilson Foundation made one of the largest philanthropic gifts in American history, devoting $200 million in support of legacy parks and greenways in western New York (Buffalo) and southeastern Michigan (Detroit). Greenways are an attractive outlet for philanthropy, as they address multiple societal issues and concerns including conservation of natural resources, equitable access to outdoor resources close to places where people live and work, environmental education, support for an active outdoor lifestyle, arts, entertainment, and love of nature.

Lasting Value of Conservation

Americans tend to value land based on what we can build on it, or how land can be mined, timbered, or otherwise exploited for its natural resources. In terms of real estate value, the 2,100 acres is worth tens of millions if it were going to be sold for houses, shopping centers, or an industrial park. The Close family determined that the highest and best use of this land was for conservation. As a landscape with intact forests, lakes, streams, and open meadows, the conserved landscape is of even greater value to area residents. The greenway improves the value of land immediately adjacent, and parcels of land situated more than a mile from its borders. The greenway is one of the largest undeveloped tracts of land remaining in the region and has become a central park for the surrounding suburbanized areas.

The presence and proximity of the greenway has influenced the pattern of surrounding land development. The greenway serves not only as a park for nearby residents; it also influenced the size of lots in adjacent residential subdivisions, the need for larger individual greenspace within those lots, and the types of amenities that homeowners seek when purchasing a lot. In the case of this greenway, adjacent land has been developed using smaller lots, a more compact suburban form, and a smaller overall land development footprint.

The gift of the Close family reinforces the importance of land and water conservation and the need for American communities to balance land development with the preservation of natural landscapes and native ecosystems. Conservation of land and water is important to the health and wellness of our communities, protects our limited supply of potable water, conserves the native habitat of plants and animals, absorbs excess rainwater mitigating impacts from flooding, and limits the amount of erosion of native soils. Conservation is a choice that landowners make. The Close family dedicated approximately one-third or 30 percent of the original family estate to conservation. That ratio is an excellent model for others to emulate.

2

Come Hell and High Water

Greater Grand Forks Greenway, Grand Forks, North Dakota

Runners participate in a race along a portion of the East Grand Forks Greenway. Source: Greater Grand Forks Convention and Visitors Bureau.

"MR. FLINK," THE CALLER BEGAN, "This is Karen Nagengast. I'm a landscape architect with the U.S. Army Corps of Engineers, St. Paul District," she continued. "We are working on the flood recovery for Grand Forks, North Dakota and East Grand Forks, Minnesota. We have been asked by the Congressional Delegation of North Dakota to solve the Red River flooding problem using 'greenways' as a solution. We purchased and read your book [*Greenways: A Guide to Planning, Design, and Development*, published in 1993 by Island Press] and want you to join us for a workshop in Grand Forks. We are hoping that you can assist us in crafting a greenway plan for the communities."

It was the fall of 1997. I had followed the flood event in Grand Forks with interest. I was preoccupied at the time of Karen's call with another client. The City of Louisville and much of Jefferson County, Kentucky, had also been devastated by a historic flood event in the spring of 1997, one month earlier than the Red River flood. My client, the Louisville and Jefferson County Metropolitan Sewer District (MSD), and specifically executive director Gordon Garner, had in essence given me a blank check and asked me to assemble the best flood recovery team in the nation, which I did immediately. I was under contract with MSD to craft a greenway plan for Louisville and Jefferson County, and as part of that contract we were attempting to update the flood protection ordinance for the county when the Louisville flood occurred.

Louisville and Jefferson County, Kentucky, suffered less significant damage, by comparison, than Grand Forks and East Grand Forks. In Louisville, flood waters damaged 40,000 homes, and the economic loss exceeded $20 million. In Grand Forks and East Grand Forks, the "Flood of the Century" destroyed both communities, above and below ground. Grand Forks and East Grand Forks evacuated an estimated 52,000 residents. Portions of downtown Grand Forks caught fire and eleven buildings burned to the flood water line. It took more than $2 billion in federal aid and almost a decade to fully rebuild the communities. It was one of the worst natural disasters in the history of the United States, a full decade before Hurricane Katrina struck New Orleans and the Gulf coast states. So when Karen invited me to Grand Forks, I was already keenly familiar with flood events, the devastating effects that flooding has on local communities, and the flood recovery process.

"Come Hell and High Water" was a phrase coined by *Grand Forks Herald* publisher Mike Maidenberg and editor Mike Jacobs in April 1997. The slogan succinctly described the total devastation that the flood had on both communities. It became the unofficial motto of the newspaper, which symbolically rebuilt its burned-out offices in downtown Grand Forks after floodwaters receded. With the Army Corps of Engineers invitation in hand it was going to be my job to convince skeptical citizens, community leaders, North Dakota and Minnesota officials, and the federal government, that a greenway was both an integral and essential element of the rebuilding process, as well as a permanent solution to the repetitive Red River flooding. It would take more than three years to

complete the master plan and almost a decade for skeptics to fully understand the benefit that the greenway bestowed on their communities.

The Red River of the North is unique in North America in that it flows
from south to north. Originating at the confluence of Bois de Sioux and
the Otter River in South Dakota, the Red River flows into Lake Winnipeg
in Canada. The south to north flow means that snow melt and spring
rains in the southern part of the watershed flow into frozen river channel sections farther north, which serves to exacerbate flooding. The portion of the Red River included within the greenway study area is roughly
seven river miles in length. In the downtown area of Grand Forks and
East Grand Forks, the Red River is joined by the Red Lake River, creating
a wide expanse of river floodplain.

A New Solution to Flood Control

My father, Richard Flink, grew up in Osnabrock, North Dakota, about
eighty miles northwest of Grand Forks. Of Swedish and German descent,
the Flink family arrived in North Dakota in 1892, and endured harsh
winters and hot summers as wheat farmers on the endless plains and
prairies of the upper Midwest. It is commonplace in North Dakota for
temperatures to drop 60 degrees below zero in the winter and exceed
100 degrees in the summer. North Dakotans are hardy people. My Dad
knew all about the Red River of the North in Grand Forks and the annual
spring floods, having grown up there and subsequently graduated with
an industrial engineering degree from the University of North Dakota.
"They used to dynamite the river in the spring to break up the ice flows
and minimize the flooding," Dad recalled as we discussed the details of
the 1997 flood. "They can't do that anymore, Dad; modern environmental
laws don't offer that as a solution," I responded. I could detect that even
my father was initially a bit skeptical of the value that a greenway would
play in resolving Red River flooding.

Indeed, the Army Corps of Engineers needed a modern, innovative,
and successful approach to resolve flooding. The Corps had already made
its normal practice offer to the communities, which included $100 mil-
lion that would fund improvements to existing floodwalls and for the
construction of a river bypass channel to circumnavigate the communities. Grand Forks and East Grand Forks rejected the initial Corps offer.

The Corps would be required to think outside the box and implement a greenway-oriented solution that would define a long-term, sustainable solution for the frequently occurring spring floods. For these communities the Corps was being held to a new standard, by Congressman Byron Dorgan, that was not normal practice. It caused the Army Corps significant concern, as the Corps uses a cost-to-benefit scenario for each flood recovery project. Costs were an even greater concern for the local communities, who would be responsible for building, maintaining, and operating a newly created greenway.

Karen and her colleague John Fisher, also a landscape architect, along with a cost engineer had put together a first draft greenway concept in July 1997, in response to Congressman Dorgan's request, after having vetted the concept with the chain of command in the Corps St. Paul District offices. The Corps concluded that a greenway solution could be part of the flood recovery program for Grand Forks and East Grand Forks as long as the local communities would provide 50 percent of the funding for this "betterment." The July 1997 conceptual greenway plan was short on specifics and did not define how the greenway would be funded. It did, however, determine a rough cost estimate for greenway facilities, including trails, a boat launch for the Red River and other amenities such as benches and trash receptacles. The total cost for this greenway betterment was estimated at $4 million.

It is not totally clear to me what transpired between the summer of 1997 and later that fall, but it was clear in my conversation with Karen in November that much more detailed planning, design, and financial work would have to be undertaken if the greenway would be included as an integral component of the flood recovery program. She also conveyed the need for me to assist the Corps in justifying the added expense and benefit of the greenway. The landscape architects within the St. Paul District offices understood the long-term benefit of the greenway and were advocates for the project. But landscape architects were not in charge of the flood recovery effort. That was the responsibility of project engineers. They were the ones that needed to better understand the greenway cost-to-benefit ratio.

To further complicate the cost-benefit requirement, as one element of the flood recovery process, the Corps and the Federal Emergency Management Agency (FEMA) would condemn and permanently remove

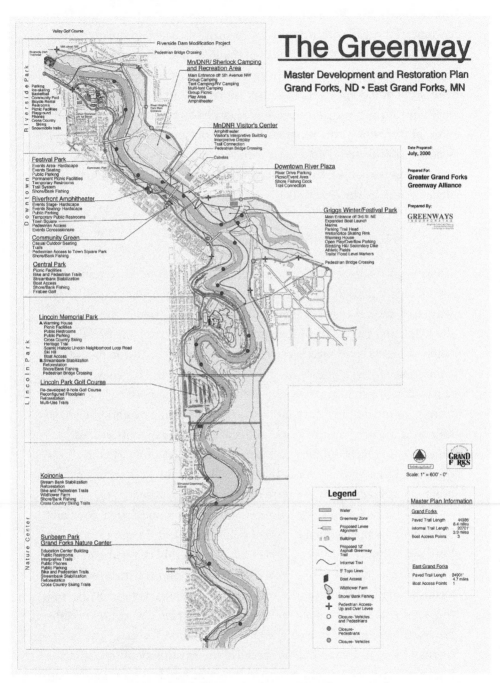

Greater Grand Forks Greenway System Map. Source: Author.

more than 500 homes from some of the historic residential neighborhoods bordering the river, as well as businesses and restaurants from commercial areas within both downtowns, in order to create a high-water flood channel on both sides of the Red River. The expanded flood channel would have to be wide enough to convey future flood events. The 1997 flood event crested an astonishing fifty-four feet above the normal flow of the river (flood stage is regarded as twenty-eight feet above normal river flow). The process of carving out the flood channel, which would be bordered on both sides with very tall floodwalls and dikes, created a broad expanse of unused land between the two cities. This land, by default, became the greenway.

The Psychology of Flooded Communities

It has been my experience and observation that flooded communities are as devastated emotionally as they are damaged physically. I have witnessed several stages of emotional distress that flooded residents endure. The first stage is the flood fight. This occurs during the flood event, when residents band together to fill sandbags and work throughout their community to save homes, neighborhoods, schools, churches, grocery stores, downtown businesses, and libraries from floodwaters. Flood fight brings out the best in a community: neighbor helping neighbor; teenage kids working arm-in-arm to fill and stack sandbags; political foes putting aside entrenched differences to save their community. Heroic efforts occur, and those communities that win a flood fight often form a strong bond that carries over into future community endeavors.

A second major stage of emotion is frustration and anger, normally brought on by the fact that the flood fight was lost, and a community has been significantly damaged. This results in a highly charged emotional environment that sweeps across a community and can turn neighbor against neighbor, dividing the community. The anger phase brings out the worst characteristics of residents. A blame game ensues. The weatherperson is to blame due to a faulty precipitation forecast. The Army Corps of Engineers is to blame because the existing floodwall was not high enough or secure enough to withstand the flood. The local politician is to blame because not enough resources were dedicated to prevent the flood in the first place. Frustration dominates community conversation, and virtually everyone wants to fix everything immediately and cannot understand why the recovery process takes so long. Irrational thought and behavior is the by-product. In Grand Forks, for example, residents of the historic Lincoln Park neighborhood could not understand why, more than a year after the flood event, they were not going to be able to move back into their flooded homes. Never mind that floodwater poured in through second story windows or, in some cases, totally submerged two-story homes. Residents of flooded communities have very short memories. They demand to know why they cannot return to their homes and neighborhoods.

A third major emotion of flooded communities is resignation and despair. People literally give up the fight, are emotionally spent because of outrage and anger, and feel lost and bitter about virtually everything associated with their community and the rebuilding efforts. For example, studies of the Hurricane Katrina disaster in New Orleans determined that suicide rates spike within flood-ravaged communities; the disillusionment and despair can become that pronounced and severe. In Grand Forks, domestic violence increased by more than 24 percent a year after the flood. In Grand Forks and East Grand Forks, some residents who experienced multiple flood events simply left town never to return. In fact, studies by the University of North Dakota confirmed that an estimated 2,000 residents, roughly 4 percent of the population, permanently left Grand Forks after the flood event.

Doubt and Skepticism Abound

My first official visit to Grand Forks in February 1998 was memorable for many reasons. First was the meeting that took place among local, state, and federal stakeholders. Second, the scale and scope of the proposed greenway was 2,200 acres in size, nearly three times the size of Central Park in New York City. Third, the weather was unlike anything I had personally witnessed, as daytime temperatures barely made it to zero Fahrenheit and nighttime temperatures plunged to 40 below zero. As a Southerner, I was ill-equipped to handle the extreme cold.

The Army Corps held a two-day workshop to envision and define the greenway and the role that it would have in the flood recovery process. Karen and I had worked for a couple of months in advance of the workshop to prepare a greenway proposal for stakeholder review. I was under contract to the Corps, and it was my job to describe the benefits of a greenway and to facilitate the workshop. I was brought into the flood recovery process just as the anger phase was building throughout both communities. I was dealing with a weary flood recovery team. The greenway concept had been included on flood recovery maps in the summer of 1997 by the Chicago-based community planning team Camiros as undefined green space between the two communities along the Red River. Camiros provided no form, program, or function to the greenway concept. They were dealing with downtown Grand Forks reconstruction and revitalization. Across the Red River, East Grand Forks was working in partnership with the Center for Community Design at the University of Minnesota to develop a fifty-year master plan for the community. The energy of both communities was totally vested in these efforts. Little thought had been given to define or understand the role that a greenway would play in flood recovery and community revitalization. Skepticism about the need for a greenway ruled the two-day Army Corps workshop.

On arrival in Grand Forks, I was immediately introduced to Grand Forks Mayor Patricia Owens and East Grand Forks Mayor Lynn Strauss. Mayor Owens is a diminutive woman with a feisty, can-do attitude. What she lacked in size she made up for with heart and passion for her community. Pat loved Grand Forks (she is retired today and living in Florida) and had deep roots in the community, and my private conversations with her revealed how deeply devastated she was by the flood. In the days immediately following the flood, Pat would be seen standing with President Clinton as the federal disaster team began the flood recovery process. Much to her credit, Pat worked long hours for the benefit of her community. She secured more than $172 million in Community Development Block Grant money to help Grand Forks rebuild its downtown and residential neighborhoods. She also worked with her congressional delegation and the federal government to secure more than $1 billion in funding for buyouts and relocations of homes, businesses, and schools and for infrastructure repair. In 1999, she was rewarded for her efforts by being voted out of office by residents, who felt that she was somehow to blame for the flood and ineffective in the recovery process—proving the age-old adage that no good deed goes unpunished.

Mayor Strauss is a very different person. A stocky, gregarious man with a huge smile and a cavalier attitude, he not only survived the flood event but emerged as an important regional leader as well. Lynn Strauss was quickly able to define a progressive vision for East Grand Forks and survived the political turmoil and emotions caused by the flood to direct the recovery and rebuilding process during the following two decades. He helped transform East Grand Forks into a vibrant riverfront community.

As our two-day event was drawing to a close, and Karen, John Fisher, and I were presenting the final greenway concept to workshop participants, a signature moment occurred when a North Dakota State Parks official raised his hand and stated: "You and the Army Corps are proposing a 2,000-acre greenway in Grand Forks and East Grand Forks; that's more park land than in all of North Dakota combined," to which Mayor Owens immediately stood and responded: "How will our small communities take care of all of this parkland? We don't have the resources!"

This emotional response and reaction to the greenway concept was the tip of the iceberg. It was the first of many challenges that proponents of the greenway would have to overcome in the months ahead. I understood these emotions, the anger, the frustration, the indignation. How could anyone propose something as frivolous as a greenway when residents were homeless and living on cots in federal and state shelters? Why should the community shoulder the burden of such a large natural, park-like landscape, when every road in the community was being ripped up so that new water and sewer lines could be reconstructed? How brazen was it to propose a brand-new community endeavor when people were out of work, the local economy was in shambles, and life in the community was anything but normal?

Clearly, the path ahead for a successful greenway project was going to be extremely challenging. This project was not simply about aligning a trail along a river or expanding an existing park and recreation program so that children could ride their bikes and roller skate, nor was it solely about conserving and connecting local resources together to improve habitat for wildlife. The proposed Red River Greenway was an enormous swath of land that would, in the years to come, be situated behind a fortresslike flood protection system. It could easily have become a colossal failure and tremendous economic and social burden to the two small neighboring communities situated in a part of North America that many Americans have never heard of or visited. Everyone involved with the greenway was exposed: the congressman who lobbied for the idea; the Army Corps who reluctantly agreed to include the greenway as an element of the flood recovery program; the community leaders who rightfully felt that they had no choice in the matter; and the guy from North Carolina, the so-called expert, who somehow, magically was going to define a program of action for 2,200 acres of leftover land that was to be transformed into a successful community betterment project. The odds were stacked against us.

Congratulations, You're Hired

In April 1998, almost a year to the day of the Flood of the Century, I presented the cities of Grand Forks and East Grand Forks with a more refined and updated greenway conceptual plan, which defined an ambitious and forward-looking vision:

> The Red and Red Lake Rivers Greenway will protect residents of Grand Forks and East Grand Forks from flooding, provide opportunities for economic growth, improve and restore ecological stability of the river corridor, link residents and tourists to four seasons of recreation and transportation facilities, provide linkage between the cities, preserve and promote the history and culture of the region through education, and improve the quality of life for future generations.

The conceptual plan defined a strategy for the continued planning, design, and development of the greenway. Goals and objectives were established, greenway management strategies were laid out, a plan for funding the greenway was described, and a five-year action plan was provided. My work as a consultant to the Army Corps was complete. The Corps would fold the greenway recommendations into a larger flood recovery plan and eventually obtained approval for the greenway from the communities in July 1998. With that approval came the next steps. One of the key recommendations in the conceptual plan was the need for a thorough and comprehensive master plan for the greenway. This subsequent

Runners enjoy a sunny day on the Grand Forks Greenway. Source: Greater Grand Forks Convention and Visitors Bureau.

plan would take the recommendations provided in the conceptual plan to a more detailed level of planning, design, and implementation.

In October of 1998, my company entered into a contract with the City of Grand Forks to begin the process of preparing the design development master plan for the Red River Greenway. City engineer Charles Grotte, greenway coordinator Sarah Hellekson, landscape architect Helen Cozzetto, and later new greenway coordinator Melanie Parvey-Biby made significant contributions to the master plan. I also worked with Greg Ingraham and Associates, a Minneapolis-based landscape architecture firm, to incorporate recommendations for East Grand Forks into a master plan for the Red River Greenway as a whole.

The Greenway Master Plan was very detailed and took approximately two years to complete. The eighty-two-page action plan report included an extensive evaluation of existing conditions for the 2,200-acre area of land, encompassing the real estate in both Minnesota and North Dakota. We expanded the greenway vision statement to include goals and objectives for the project. The plan described the type of greenway facilities that would be constructed, including an amphitheater, all season trails, festival areas, a community green, boat access, and wildlife viewing areas.

My team provided design standards for specific facilities to guide future development and construction, a greenway operations and maintenance program, as well as a detailed funding strategy for building and managing the greenway. The final chapter of the plan provided a step-by-step program for implementing the recommendations, including a strategy for marketing the greenway to residents and visitors.

The master plan included a detailed set of maps that provided a design program for major greenway elements, such as the riverfront amphitheater, community green, proposed Grand Forks Nature Center, and refurbished Riverside Park. A map illustrated greenway management responsibility assigned to the City of Grand Forks, City of East Grand Forks, and State of Minnesota.

An Alliance of Greenway Interests

As we began the master plan process in the fall of 1998, Mayors Owens and Strauss formed a Greenway Alliance, a collection of twenty key stakeholders that helped guide the master plan decision-making process. The Greenway Alliance included representatives from both communities, officials from state and federal agencies, businessmen, university representatives, and school officials. The Alliance was charged with two important tasks: 1) assist in preparing the master plan, and 2) determine an operation and management structure for the greenway. The Alliance met monthly during 1999 to prepare the bulk of the master plan recommendations.

Many of the sustaining ideas that now comprise the Red River Greenway came from members of the Alliance. Roger Hollovet, who served as chair of the Alliance, was director of the U.S. Fish and Wildlife Service's Devils Lake Wetland Management District. Roger contributed significantly to the concept of the greenway as an important wildlife corridor. He was quick to point out that the Red River is an important ecosystem along the North American flyway for millions of migrating birds. He also shared that bird watching is a multibillion-dollar industry in the United States, and that the two communities were ideally situated to reap economic benefits by transforming portions of the greenway into a wildlife sanctuary. Roger even proposed establishing a Grand Forks Nature

Center with support from the U.S. Fish and Wildlife Service. That idea was later squashed by North Dakota Governor Ed Schafer when it was better understood that a nature center would impose restrictions to hunting and farming across the region to improve habitat for migrating birds.

John Winter, regional manager for the Minnesota Department of Natural Resources (MNDNR), was one of the biggest allies and supporters of the greenway and an active member of the Greenway Alliance. John was charged by the State of Minnesota to work directly with the City of East Grand Forks to establish a State Recreation Area within the city limits. John and his team at MNDNR were excited by the prospects of having 1,200 acres of land to work with on the Red River. John worked closely with the Minnesota state legislature to garner the funding, planning, and design assistance to East Grand Forks and the successful State Recreation Area.

Dale Skyberg and Steve Mullally were also very important to the success of the greenway. Dale was the Parks and Recreation Director for East Grand Forks, while Steve was the park superintendent for the Grand Forks Parks District. Both these men shouldered a lot of responsibility for coordinating the implementation of the greenway within their respective organizations. It was an incredibly challenging time for both of their agencies, as most park facilities were totally destroyed by the 1997 flood, and there was a need and an expectation that recreation facilities and programs would be returned to functional operation in short order. To complete their duties, they had to incorporate a 2,200-acre greenway into their work plan and agency responsibility.

The Alliance worked diligently for more than a year on the master plan, meeting monthly and determining key components of the master plan. The formation of the Alliance was so successful that it continues to this day in an advisory capacity for both communities.

It's Always about the Economy

The *Grand Forks Herald*, led by Maidenberg, Jacobs, and Tom Dennis, remained a steadfast supporter of the greenway. The *Herald* staff effectively utilized their resources to keep the greenway concept front and center in the community discussion. Despite the lack of early community support,

Maidenberg and Jacobs kept publishing articles and editorials that positively depicted the greenway. They also regularly chastised the leadership in Grand Forks for being slow to act on finalizing the greenway master plan, and they cheered East Grand Forks as Mayor Strauss deftly moved from concept to implementation.

In early 2000, Michael Brown was elected mayor of Grand Forks. I never had the opportunity to get to know Mayor Brown the way that I grew to know Mayor Owens. What I quickly learned was that first and foremost Mike Brown thought like a businessman and not a politician. He quickly grasped the potential of the greenway and supported the conclusions and recommendations of the master plan. It was also clear to me, through my discussions with Mayor Brown, that financing and funding the greenway was of paramount concern.

Across the Red River, Mayor Strauss was fast at work. The City of East Grand Forks was determined to land a major retailer to anchor their redeveloped downtown. They wanted a name brand retailer that would increase resident and visitor traffic. The city worked to recruit several retailers, and their hard work paid off in December 1998 when Cabela's Incorporated decided to locate a storefront in downtown East Grand Forks. Cabela's is known as the world's foremost outfitter. They proposed developing a 60,000-square-foot, five-story superstore smack in the middle of East Grand Forks, Minnesota—with a resident population of 8,000. When Mayor Strauss asked Cabela's how many visitors the store would attract annually, the company responded "about one million." The mayor worked quickly to designate and build enough parking spaces to accommodate the projected influx of store patrons. This would be the fifth Cabela's store in the United States and the first to be located in a downtown. When asked about the unique location of the Cabela's store, company representatives cited the location of the greenway as the primary motivating factor. After all, representatives reasoned, the company sells outdoor equipment, clothing and accessories, so what better place to immediately try out your new purchases than within one of the largest urban greenways in the Midwest. Downtown East Grand Forks was a perfect fit for the new Cabela's store.

The Cabela's announcement set wheels in motion, most importantly that a major retail company would base their growth and expansion plans on the fact that a 2,200-acre greenway would, someday, exist between the cities of East Grand Forks and Grand Forks. At the time, the reality of that future greenway was more uncertain than the completion date for store construction and its grand opening.

In the spring of 1999, six months into the operation of the store, I visited with the Cabela's store manager to get a first-hand account of his operations. He immediately confessed that the company had missed the mark on actual visits to the store. I responded that it was totally understandable, and that their projection of a million store visitors seemed a bit excessive to me. He quickly shot back, "yeah, we are well on our way to having two million visitors this year." Stunned, I asked him to clarify why the store was so popular. He said, "I don't know, probably our location here on the greenway. It doesn't much matter to us anyway, because most of the company sales are generated through our catalog."

In early 2000, the greenway master planning process had reached the stage where important financial matters had to be defined and resolved. It was clear that the master plan would finally be able to answer the nagging question posed in the winter of 1998 as to how the communities would care for and manage the greenway. Up until this point in the process, the question could not have been answered because the master plan team did not have a defined development and operations program for the greenway. After two years of community discussion, intense planning work, and sorting through a variety of multijurisdictional issues and project constraints, the time had arrived to tackle the financial questions head on.

At the time, the City of Grand Forks leadership was virtually paralyzed by the thought of paying for the cost of operating and managing the greenway. I proposed to both communities an independent multijurisdictional organization to lead the operations and management program. The cities and the Grand Forks Park District, however, decided to keep operational control within their institutional framework. I shared with City Council a financial analysis that defined how the greenway could generate revenue and thereby support a portion of associated operating costs. I convened meetings with park and recreation officials and nonprofit groups to discuss event programming for the greenway. After several meetings an agreed-upon schedule of monthly events emerged for all four seasons. Some of these events would charge an admission fee. It was a shared belief that the greenway events would draw visitors from outside the Greater Grand Forks community.

In late July of 2000, the master plan was nearly complete, and a critically important meeting had been scheduled. The entire flood recovery and greenway team gathered at City Hall to present a coordinated presentation to the Grand Forks City Council Flood Protection Committee. In hindsight, it was one of the most important presentations of the master planning process, which is not to say that the meeting went well.

I presented a summary of my conservative financial proposal to define how the greenway could generate revenue to pay for portions of greenway operating costs. My presentation, titled "Greater Grand Forks Greenway: An Estimate of Potential Revenue" was designed to put to rest the concern that the greenway was destined to become a financial burden on the local communities. My goal was to change the perception from "this is going to cost you a lot of money" to "the greenway might pay for itself." The presentation included facts and figures on ways in which the greenway would generate revenues from the hosting of a few local events. I estimated that more than 200,000 tourists would use the greenway each year and calculated how much the average tourist might spend in the community on things like hotel stays and dining, as well as estimates of ticket sales and concession receipts. I then described direct and indirect revenues from these expenditures and concluded by stating that from this conservative hypothesis, the total annual economic benefit of the fully developed greenway could be defined as $16 million.

Some members of the Council subcommittee exploded with anger and disbelief. One of the Council members angrily challenged my conclusions, stating "Mr. Flink, do you know where you are right now? You are in Grand Forks, North Dakota. You are in the middle of NOWHERE, our community is on the way to NOWHERE, and NOBODY is going to visit your damn greenway." A few committee members seemed convinced that I was a snake oil salesman. How could anyone possibly believe that the greenway would attract tourists and generate revenue? How could someone have arrived at such preposterous conclusions concerning revenue generation from a greenway? It was total hogwash, and certain committee members would have none of it.

John Winter was in the audience that evening. He could see that I was struggling to convey my proposal to City Council. He stood and addressed the Council and calmly stated that, first of all, he was totally unaware that I had prepared a financial model for the proposed greenway, and second, the Minnesota legislature had asked his department to complete a similar study, and third, amazingly, the results of his study were similar to my conclusions. The greenway could generate net revenue from events and therefore offset operational costs. At that moment, I thought that some members of the committee were going into cardiac arrest. They were irritated and angry with both of us. In total frustration I responded that the City didn't need to believe either John or me, but it would be in the best interests of the community to hire an independent financial adviser to determine the revenue potential from greenway events and its impact on operational costs.

Maidenberg and Jacobs couldn't resist the hanging curve ball that John Winter and I launched into the air that evening. The morning headline in the *Herald* boldly proclaimed: "Consultant: Greenway can be a money maker. Recreation area could generate $16 million annually." The following week, a Dennis editorial in the *Herald*, titled "Grand Forks Greenway: An eagle, not an albatross," correctly described the fact that the "city leaders seem strangely reluctant to look at the project with enthusiasm." Dennis concluded, "Grand Forks can enjoy its own Greenway, or it can enjoy a view of the Greenway in East Grand Forks. At this point, energy, not money, is the key. Because a Greenway is a realistic public-policy objective. It is not an impossible dream." Later that fall, a Jacobs editorial pleaded with the City Council—"Grand Forks should back a plan for the greenway." Jacobs concluded his editorial stating, "First, Grand Forks needs to find a way to make the greenway happen on the North Dakota side of the Red. Exactly what's necessary isn't entirely clear at this moment, but a commitment that something must happen will go a long ways toward finding the means to make it happen."

At this point in the master plan process, my frustration with the leadership in Grand Forks and lack of comprehension regarding the benefits of the proposed greenway was reaching a crescendo. I wasn't the only one who was frustrated. I made presentations to an enthusiastic Grand Forks Chamber of Commerce, the Rotary Club, and other key community stakeholders, all of whom questioned why there was such lack of support for the greenway. East Grand Forks, meanwhile, continued to move ahead with speed and conviction, and had begun design work on several components of their greenway. On the west side of the Red River, the squabbling, indecision, and lack of direction was a drag on the Grand Forks efforts.

In hindsight, it is hard to know exactly when the light bulb went on in the minds of the Grand Forks leadership. I honestly don't believe that that moment ever occurred. It is understandable given everything that the Mayor and City Council had on their plate at the time. There was an official groundbreaking ceremony for the flood recovery program in June of 2000, and at the same time an official groundbreaking took place for the greenway. However, it would take more than a year from the date of my July 2000 presentation to the acceptance and adoption of the final greenway master plan in September 2001. By this time, I was disillusioned about the four-year planning experience and was unsure that anything would ever happen with the greenway in Grand Forks. I remained enthusiastic about the recommendations in the master plan and was hopeful that the city would abide by the detailed strategies provided.

It is fair to conclude that inclusion of the greenway within the Army Corps of Engineers flood protection program was probably the most important factor in making the greenway project a reality in Grand Forks. The future success of the greenway was not the result of a community champion or civic leadership. Most of the community leaders in Grand Forks were less than enthusiastic about the project. So the project was turned over to the Army Corps and a team of landscape architects and engineers, who, piece-by-piece, landscape-by-landscape, brought the greenway to life.

Who Could Have Ever Imagined . . .

My intimate involvement with the Greater Grand Forks Greenway (also known as the Red River Greenway) concluded in the fall of 2001. Much of the success of the greenway, from a facility design and construction point of view is, in my opinion, due to the outstanding work of landscape architect Tom Whitlock and the team of Stanley Engineers and Damon Farber Associates (DFA). The Stanley/DFA team was retained by the Army Corps of Engineers to execute the recommendations of the master plan. Tom and his team faithfully implemented all the key elements defined in the adopted greenway master plan. Furthermore, as evidenced by the completed greenway today, their design work is extraordinarily beautiful and exceptionally functional.

When you visit the Red River Greenway today, none of the past challenges associated with the planning, design, and development are evident. All you will encounter are smiling faces, engaged users, happy merchants, and enthusiastic local officials. In the city of Grand Forks, the Rotary Park, Town Square, Community Green, and Festival Lawn Amphitheater are fully functional and host a variety of public events throughout the year. The transformation of the historic Lincoln Drive residential neighborhood into a community park was a tremendous challenge and today is a beloved, if not sacred landscape. Beautifully designed floodwall gateways provide a seamless transition through the towering levees, linking residents to the greenway.

The total costs of greenway facility construction, regarded as a betterment by the Army Corps of Engineers, exceeded $20 million, which is much higher than the Corps original estimate of $4 million defined in their July 1997 concept plan. However, it is a fraction of the more than $2 billion that was spent on the entire flood recovery project.

Bridges across the Red River link the two communities, including the Pat Owens Bridge, named for the former mayor. The twenty miles of trail network extends across the entire greenway, and in 2007 it was designated a part of the National Heritage Trails system.

On the East Grand Forks side, the Red River State Recreation Area encompasses more than 1,200 acres of land and is owned and operated by the Minnesota Department of Natural Resources. The state park provides full-service camping in an area that was once the Sherlock Neighborhood, which was removed under the FEMA buyout program after the flood. East Grand Forks also proudly boasts The Boardwalk, which is lined with restaurants and bars, including the Blue Moose Bar and Grill, and historic Whitey's Café. An invisible floodwall in the downtown offers East Grand Forks unobstructed views into the greenway for many months of the year. Cabela's continues to be a flagship store in the heart of the community.

One of the first major events that incorporated elements of the completed greenway occurred in June 2004. The first Greenway Day in Grand Forks and East Grand Forks attracted more than 50,000 participants; so

A family walks along a portion of the Grand Forks Greenway. Source: Greater Grand Forks Convention and Visitors Bureau.

much for being "in the middle of nowhere" and "on the way to nowhere." Since that first greenway event, the program of events envisioned in the master plan has matured and expanded to meet the local demand and opportunity for regional events. Each February, Winter Fest is a substantial and successful production that showcases the unique assets of the frozen Red River valley. Each summer, the Grand Cities Art Fest is one of the most important regional events held in the upper Midwest. In addition to these events, Catfish Days, the Chili Cook-Off, and the Potato Bowl are local events that thrive within the greenway. The idea that the greenway could support events and generate revenue has come true.

The recovery and redevelopment of Grand Forks and East Grand Forks is a national success story. Both communities have invested hundreds of millions in public and private development. Grand Forks is a self-proclaimed Destination City, I believe in no small measure due to the success of the greenway. In February 2006, the *New York Times* ran a feature article in the travel section of the newspaper titled "36 hours in Grand Forks." It features a wintertime trip to the upper Midwest city. The article lets tourists know how to travel to Grand Forks, where to stay, and what to do, and the greenway is prominently featured in that article.

In fact, the photo that accompanies the article is a beautiful image of a cross-country skier using the greenway.

In May 2006, *Fortune Small Business* magazine ran an extensive feature on the communities, titled "River Revival: Grand Forks, N.D. has come a long way back since a disastrous flood in 1997, could it teach New Orleans a thing or two?" The magazine profiled the new businesses that have opened in both communities since the flood, including the Amazon.com call center, employing 300, and LM Glasfiber, which employs 300 in an expanded 150,000-square-foot manufacturing facility.

It is fair and appropriate to conclude that the greenway is very much an integral part of life in both communities. Residents enjoy year-round activities, community events, recreational and educational programs, and a wide variety of outdoor activities. The greenway has become the centerpiece of an active outdoor lifestyle. A summary of community attitude surveys on the Grand Forks web page (http://www.greenwayggf.com) and Facebook page (The Greenway of Greater Grand Forks) supports the fact that the greenway may be the most beloved landscape in the Greater Grand Forks area. Community residents are quick to point out how important the greenway has become to the quality of life in both communities.

Cross-country skiers enjoy groomed trails on the Grand Forks Greenway. Source: Greater Grand Forks Convention and Visitors Bureau.

THE GREENWAY
GRAND FORKS / EAST GRAND FORKS

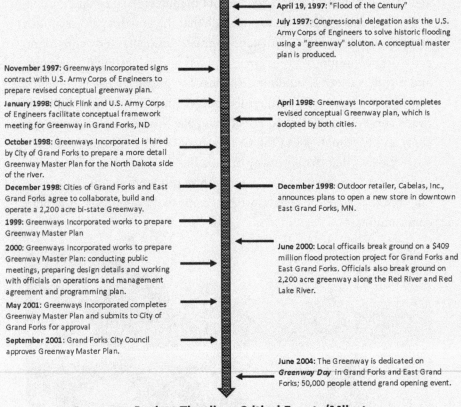

April 19, 1997: "Flood of the Century"

July 1997: Congressional delegation asks the U.S. Army Corps of Engineers to solve historic flooding using a "greenway" soluton. A conceptual master plan is produced.

November 1997: Greenways Incorporated signs contract with U.S. Army Corps of Engineers to prepare revised conceptual greenway plan.

January 1998: Chuck Flink and U.S. Army Corps of Engineers facilitate conceptual framework meeting for Greenway in Grand Forks, ND

April 1998: Greenways Incorporated completes revised conceptual Greenway plan, which is adopted by both cities.

October 1998: Greenways Incorporated is hired by City of Grand Forks to prepare a more detail Greenway Master Plan for the North Dakota side of the river.

December 1998: Cities of Grand Forks and East Grand Forks agree to collaborate, build and operate a 2,200 acre bi-state Greenway.

December 1998: Outdoor retailer, Cabelas, Inc., announces plans to open a new store in downtown East Grand Forks, MN.

1999: Greenways Incorporated works to prepare Greenway Master Plan

2000: Greenways Incorporated works to prepare Greenway Master Plan: conducting public meetings, preparing design details and working with officials on operations and management agreement and programming plan.

June 2000: Local officails break ground on a $409 million flood protection project for Grand Forks and East Grand Forks. Officials also break ground on 2,200 acre greenway along the Red River and Red Lake River.

May 2001: Greenways Incorporated completes Greenway Master Plan and submits to City of Grand Forks for approval

September 2001: Grand Forks City Council approves Greenway Master Plan.

June 2004: The Greenway is dedicated on *Greenway Day* in Grand Forks and East Grand Forks; 50,000 people attend grand opening event.

Greenway Project Timeline: Critical Events/Milestones

A timeline illustrates portions of the greenway planning and design process. Source: Author.

Why Are We Experiencing More Flood Events?

The problem is twofold: (1) we are experiencing more extreme weather events as a result of a warming planet, and (2) more people are moving into hazardous landscapes in unprecedented numbers. The extreme weather events are the result of more water vapor in the atmosphere, which is a direct result of a warmer planet and melting of land and sea ice at Earth's polar regions. As the hydrologic cycle dictates, what goes up must come down. In this case more water vapor means more rainfall. Also, wet moist air is unstable, while dry cool air tends to be more stable. More water vapor provides the fuel necessary to transform small storm events into supersized storm events.

Simultaneously, the world's population is moving into shoreline communities. For example, in 2003, three billion people lived within 200 kilometers of a shoreline. By the year 2025, that number will double to six billion people.[1] In the United States, 39 percent of all residents, approximately 123 million people, live in counties that border a shoreline. It is expected that between 2010 and 2020, that will increase by 8 percent.[2]

When you combine the facts that extreme weather events are occurring with greater frequency and at the same time more people are moving into vulnerable landscapes, the result is outsized disasters that impact lives and damage property. That is what we are regularly experiencing across the American landscape, and it is happening with greater frequency on all major continents on the planet.

A More Resilient Grand Forks

Not long after the spring 1997 flood waters had receded, the U.S. Army Corps of Engineers began to contemplate a complex question; should we move the cities of Grand Forks and East Grand Forks from their historic locations, or should we rebuild the communities in place and make them more resilient to future floods? This is a question that is asked after every major disaster. Should New Orleans be moved or rebuilt? Should parts of the Jersey Shore be moved or rebuilt? Should North Carolina's first historically African American community, Princeville, be relocated or rebuilt? It is an easy question to ask and very difficult to answer. In the case of Grand Forks and East Grand Forks, the decision was made to rebuild the communities with modifications. This decision in fact tests the theory of resiliency. The solution can't be to build back to the condition that was present before the impactful event. Modifications to community settlement must be made. More greenspace must be provided as land capable of absorbing the next storm event—which is certain to happen. Becoming a resilient community includes adapting to a dynamic environment.

Literally from the ashes and destruction of flood-ravaged communities have risen two competitive, quality-oriented, vibrant towns whose future is brighter today than on that dark and cold April day in 1997. The greenway, heart and soul of this rebirth, is the result of the collective efforts of dozens of people who, during the past decade, have helped to shape and develop a landscape that tens of thousands enjoy each year. As this story has defined, various planners, landscape architects, engineers, construction contractors, park and recreation administrators, and citizen advocates have worked from a shared vision to realize what, at the outset, seemed to many an impossible dream.

The Grand Forks Greenway is regarded by the Corps as one of the nation's best flood protection projects. "It's about the best example (in the nation) we have to date; it responds aesthetically; it responds functionally," states Kevin Holden, landscape architect with the U.S. Army Corps of Engineers, in a February 2010 *Landscape Architecture* magazine article titled "When Rivers Rise." The greenway serves its intended function, absorbing spring floodwaters on an annual basis. Several severe flood events, most notably in 2008 and 2009, have tested the capacity of the fifty-four-foot-high flood levee and protection system. The 2009 flood was the fifth highest in recorded history, with the Red River cresting at fifty-two feet on March 22 of that year; however, no damages occurred in either community due to the flood protection measures installed after the 1997 flood event.

Sadly, national practice has not built on the success that Grand Forks and East Grand Forks now enjoy. Floods continue to devastate riverfront communities across the nation, including upriver in Fargo, North Dakota. In 2008 and 2009, for example, the communities of Des Moines and Cedar Rapids, Iowa, were devastated by flood events. Officials from Cedar Rapids traveled to Grand Forks to view what the two cities accomplished. But that kind of independent effort should not be the standard for information sharing. The Army Corps of Engineers and the federal government should be more proactive and not wait until natural disasters devastate communities. Riverfront towns and cities should be tutored and educated well in advance of such disasters to prepare for inevitable damage associated with river flooding. Excellent models exist, including the Red River Greenway, the Napa River Flood Protection Project in California, and the South Platte River Greenway in suburban Denver, Colorado. New projects of similar size are emerging across the nation. For example, the City of Dallas, Texas, is proposing a project of similar scope and scale as the Red River Greenway for the Trinity River.

The story of the Greater Grand Forks Greenway illustrates the need for communities to respect the powerful forces of nature and realize that abundant green space in close proximity to our communities, especially along rivers, is an asset that can be planned and developed to protect public health, safety, and welfare, support economic prosperity, and improve the quality of life for generations to come. Adopting this approach, the United States can begin to adopt the mentality of the Netherlands, which after a horrific flood in 1953 began to embrace a lifestyle and technology that enabled them to flourish. Large communities such as Rotterdam are physically below sea level but don't flood. That is because the Dutch have invested a significant amount of capital into making their communities resilient to flooding. These are lessons that the United States should pay more attention to and begin to adapt if we are to have our shoreline communities become more resilient to frequent storm events.

Make Your Community Resilient to Floods

If you live in a shoreline community, chances are you will experience a flood. The Federal Emergency Management Agency has determined that all fifty states have been subject to life-threatening and property-loss floods during the past decade. Shoreline communities are the most vulnerable. Here are a few things you can do to prepare for this natural disaster:

Purchase Flood Insurance

Make certain that those residential homeowners and businesses that are located in mapped flood-prone landscapes have purchased and maintain their flood insurance. This is an insurance policy separate from all other property insurance that must be purchased through the National Flood Insurance Program. An insurance agent in your community can help affected property owners determine if they are eligible for the insurance.

Update Floodplain Maps

Many of the maps that are used to depict the size and extent of flood-prone landscapes are woefully out of date. Check with your municipality to verify the last time floodplain maps were updated. Work with the Federal Emergency Management Agency and the U.S. Army Corps of Engineers (USACE) to update the maps.

Map Repetitive Flood Loss Properties

Work with the private insurance market and local governments to determine properties that have been flooded multiple times. These properties are the most vulnerable to future flooding. It may be most cost effective to move these residential homes and businesses out of frequently flooded areas. Map these properties and work with FEMA and USACE to undertake steps necessary to relocate people and businesses from these landscapes.

Convert Vulnerable Properties to Greenspace

What I have learned from working on natural disaster events is that the highest and best use of flood-prone landscapes is for the storage of flood waters—and not for residential housing and businesses. The smartest and best economic decision that can be made is to ensure that these lands remain community greenspace. Work with local government planners and designers to expand public community greenspace to include those lands most vulnerable to flooding.

Prepare an Evacuation and Recovery Plan

Complete a community evacuation and recovery plan. Work with residents of your community to complete the plan, make printed copies of the plan and ensure that all residents and businesses have a copy of the plan. Communications are vital during and after natural disasters. From my experience, I have witnessed firsthand that residents and businesses don't know what to do during a flood event and after the floodwaters have receded. An evacuation and recovery plan provides essential communication and guidance to every citizen of your community.

Hire a Community Resiliency Officer

A recent trend has been to employ a Community Resiliency Officer, who can work with local government, law enforcement, and state and federal officials prior to, during, and immediately after a natural disaster. Resiliency Officers will typically focus on a broader agenda, beyond flooding. For more information visit https://100resilientcities.org/resources/.

3

Turning Trash into Trails

Swift Creek Recycled Greenway, Cary, North Carolina

Two women walk along a portion of the Swift Creek Recycled Greenway in Cary, North Carolina. Photo by author.

Sustainable Living

What does it mean to live a sustainable life? What actions must we take to live sustainably? These questions are being asked with increasing frequency by twenty-first-century Americans. As world population soars past 7.4 billion and the population of the United States exceeds 340 million, the impact of people on the resources of planet Earth is becoming relevant. To live sustainably, should we lower our consumption, use less water, commute by bicycle instead of by car, grow our own food, and lower our carbon footprint?

If we consider planet Earth as a bank account, with the soil, minerals, water, plants, animals, and air as the deposits in this account, let's realize that we are withdrawing the assets of this bank account at an accelerated rate, while depositing virtually nothing back into the account. The bank account of the Earth is being depleted, and there is justifiable concern that future generations will have less of the bank account to maintain future quality living conditions.[1]

Environmentalist and entrepreneur Paul Hawken has concluded, "Sustainability is about stabilizing the currently disruptive relationship between earth's two most complex systems—human culture and the living world."[2] In 1987, the Brundtland Commission of the United Nations declared, "Sustainable development is development that meets the needs of the present without compromising the ability of future generations to meet their own needs."[3] The commission's findings were further supported in 2005 by a World Summit recommendation that sustainability consider three overlapping demands of global society: environmental, social, and economic. This is known as the triple bottom line and is increasingly embraced as an essential metric for measuring sustainable practices. In effect, all projects, regardless of type, size, or impact, must be measured against the triple bottom line in order to understand their true impact.

There are so many questions associated with the word "sustainable" and its applicability to the American lifestyle. We are constantly reminded of the fact that we are a nation of consumers. Americans constitute 5 percent of the world's population, and yet we consume 17 percent of the world's energy.[4] Consumption seems to be synonymous with being an American, and we are the consumers that our global economy depends on to sustain economic health. But now we are being asked, told, and reminded of the need to cut back on consumption, to conserve resources, and to be more mindful of our impact on the natural resources of the planet. But how much can and should we change our lifestyle to satisfy metrics of sustainability?

Planet of Trash

We live on a planet that is rapidly filling with trash, refuse, garbage, discarded material, and other assorted human waste. This human-caused reality is the by-product of the industrial revolution and more than seventy-five years of consumerism. Our oceans are being polluted with millions of tons of plastic and other trash, as nations throughout the world either use refuge barges to intentionally dump the trash offshore or negligently allow river systems to transport carelessly discarded waste, trash, and refuse from adjacent shoreline communities. Humans on every continent are contributing to the dramatic dumping of trash into our oceans. We can all share in the blame and tragedy of this devastating by-product of civilization.

As the world's most notable consumers, Americans generate an enormous amount of trash. According to the Environmental Protection Agency, the average American generates 4.4 pounds of trash each day, 29 pounds of trash per week and 1,600 pounds of trash in a year. By age seventy-five, each American on average will have generated 52 tons of trash.[5]

Our country generates enough trash each year to bury the state of Texas three times over. The vast majority of this trash ends up in landfills, much of which could be repurposed for another functional use and a secondary life cycle. As just one example, more than 30 percent of all trash deposited in landfills is paper that is suitable for recycling and reuse. As global society has clearly embraced and promulgated industrial works and technology, it is clear that now is the time to begin thinking about ways to recycle and reuse the by-products of industrialization. We must completely rethink the cycle of raw material extraction to manufactured goods to trash.

In the landmark book *Cradle to Cradle,* architect William McDonough and scientist Michael Braungart succinctly define "cradle to grave" as one of the most significant by-products of industrialization. Cradle to grave begins with the extraction of natural resources, such as minerals, vegetation, petroleum, and water, and continues through the manufacturing process, where it is developed into a variety of usable and functional materials and products, such as chairs, plastic bottles, couches, clothing, automobiles, smart phones, and more. The vast majority of products produced in the world today have a limited singular life span, or usefulness, and most wind up being discarded and hauled off to landfills, "the grave," where they are buried or dumped and forgotten. This process of material extraction, manufacturing, and trash is not the optimal resource cycle of a sustainable society. The resulting end product is rapidly becoming one of the world's largest problems—what to do with all of the trash generated from human consumption.

Recycling Our Waste

In current economic terms, Americans place very little value on trash. As the by-product of consumption, we regard trash as a useless, messy, unsanitary item that must be discarded quickly and efficiently. Generally speaking, after we have properly disposed of our trash, usually in a trash container that is collected by sanitation workers, we tend to feel good about our actions—our job is complete. In fact, the percentage of disposed material that comprises municipal solid waste has not changed appreciably since 1960.[6] But what if trash is thought of as a resource for building new things—such as furniture, signs, fence posts, drainage pipes, and a variety of other products? What if we changed our entire mindset and no longer considered *trash* as something to discard, and we simply considered these items to be *recyclables*? Would these uses and demand for recyclables as a resource transform the way in which Americans think about postconsumption by-products and our disposal of these items? Could a better understanding of the benefits of recycling be an element of sustainable living?

Recycling and reducing waste has become a necessity for the United States. The U.S. Environmental Protection Agency reports that since 1990, municipal solid waste has leveled off and started to decline; during the same period recycling rates have risen dramatically across the nation. One of the issues that is supporting this dramatic change in attitude toward recycling is the fact that communities throughout the nation are faced with serious solid waste management problems as local landfills are burdened to capacity, and the prospect of opening new landfills becomes increasingly difficult. The solid waste recycling movement has become one of the more successful citizen-initiated environmental programs in the nation's history. Americans are slowly accepting the prospects for recyclables. Almost one-third of all waste generated in 2008 was recycled.[7] But getting Americans to purchase products made from recyclables has caught on more slowly.

The most successful products manufactured from recyclables are from paper, plastic, and metals. This is where participation among American consumers is the greatest and where recycling has had the greatest impact. For example, more than 53 percent of all paper products purchased are now made from recycled waste paper.[8]

Another form of savings is in energy consumption, where, for example, it takes 8 percent of the total energy to manufacture an aluminum can from a former aluminum can as opposed to manufacturing the same can from raw materials.[9] Recycling reduces greenhouse gas emissions. According to research completed by the American Planning Association, "recycling also helps to reduce greenhouse gas emissions that affect global climate. In 2008, the national recycling rate of 33.2 percent (83 million tons recycled) prevented the release of approximately 182 metric tons of carbon dioxide equivalent into the air—roughly the amount emitted annually by 33 million cars, or 1.3 quadrillion BTU's, saving energy equivalent to 10.2 billion gallons of gasoline."[10]

Walking the Walk

As a company with the word "green" embedded within its name, Greenways Incorporated has always been committed to sustainability. As company owner, I make an effort every day to live sustainably. I recycle virtually everything that can be recycled. I purchase products made from recycled waste. When possible, I bicycle, walk, or use transit or ride share services to reduce energy consumption. I have owned a hybrid car for more than a decade as part of my commitment to lower gasoline consumption. I am constantly aware of my carbon footprint and committed to reduce this impact. I try not to just "talk the talk," but more importantly "walk the walk."

This focus on trying to lessen my impact and live sustainably has been a part of my company since it was founded in 1986. In the spring of 1991, my staff and I began to think of ways in which common household and postindustrial trash could be turned into useful products for building greenways. We sought to demonstrate that trash could become one of the most important resources for twenty-first-century America. From this interest in recyclables emerged the nation's first greenway built from the principles and products of recycling.

The Swift Creek Recycled Greenway, located in Cary, North Carolina, is the nation's first greenway project to be constructed completely from recycled postconsumer and postindustrial waste. This remarkable award-winning project, completed in 1992, is a demonstration of the need and benefit to closing the final leg of the recycling loop—using products made from recyclables to build infrastructure of the future. The greenway is more than a recreational amenity. It serves to link commercial development, parks, and residential neighborhoods, providing an efficient nonmotorized route of travel. The greenway has been integrated into local educational curricula, teaching children the value of recycling and showing the full circle of recycling—from manufactured product to discarded waste to a new usable product.

The project was an outgrowth of regularly scheduled staff meetings at my company, where discussions focused on the ability to merge the success of the greenway movement with those of recycling. A newspaper article featuring the use of recycled rubber tires in asphalt paving got us thinking about every element of greenway trail design and construction. Terri (Kroll) Musser, one of my project managers at the time, took a strong interest in the idea of recycling and led our efforts to investigate the possibilities of building a greenway from recyclables.

A Hometown Client

With the idea of building a greenway from recyclables in hand, we needed to find a project or, more important, a client, a guinea pig of sorts, that would allow us to build our experiment. Because the concept of constructing a greenway from recyclables was leading edge, it was paramount that we maintain an intimate level of involvement with the project throughout the design development process; therefore, the project had to be close to home.

We quickly decided that the ideal location for the nation's first "recycled greenway" would need to be in our hometown, Cary, North Carolina. The town of Cary has been frequently cited as one of the best places to live, work, and raise a family. Cary is also one of the most innovative and progressive communities in the nation. Today, the town operates an exemplary graywater reuse program, recycling wastewater from homes and businesses for use in irrigation. Cary is the home of SAS, a global software company, year-after-year considered one of the best employers in the United States. Cary is one of the fastest-growing communities in North Carolina and as such faces a problem that is increasingly found in other rapidly growing American communities; its landfills are reaching capacity and the prospects for opening new ones are limited. Cary's curbside recycling program is helping to significantly reduce the amount of garbage that enters the landfill. At the time of the project, the town had achieved a remarkable 75 percent household participation in the curbside collection program. Today, Cary diverts approximately 42 percent of its waste stream from landfills.

In order to complete our experiment, we would need to draw on the town of Cary's progressive and innovative spirit. I approached an old college classmate and friend with the idea of building a recycled greenway, the town's greenway planner, Tim Brown. Tim is the type of person who loves a challenge, and his spirit of adventure is matched by a jovial and upbeat outlook on life. Tim's first reaction to the idea of building a greenway trail from trash was "Fantastic, what do we need to do to get started?" Most important, we needed a specific greenway project and ideally one that the town was ready to start building. Tim had the perfect project, the Swift Creek Greenway, located in the southeast quadrant of the community. He was almost ready to bid the project for traditional construction when he received my phone call. Excited by the idea of a recycled greenway, Tim and I then had to sell the concept to Cary's Greenway Commission, Parks and Recreation Department, and Town Council. Amazingly, with little more than an idea to work from, all parties agreed that the idea was worth pursuing. Greenways Incorporated and the town entered into a partnership agreement to transform the Swift Creek Greenway project into a recycled greenway.

The design development program for the greenway slowly emerged— like a ball of twine being unwound strand by strand. During the early stages of the project three issues became critical: (1) very few products made from recyclables were readily available locally to meet the demands for building a greenway project; (2) some of the products that we did locate and that were available had not been tested for durability, safety, longevity, and aesthetic appeal; (3) the cost of purchasing and installing recycled products, in place of those made from traditional raw materials, was excessively high. In fact, in the early 1990s the recycling movement nationwide was suffering from a failure to close the third leg of the recycling loop—convincing consumers to purchase and use products made from recyclables.

It became important for us to find products that could meet the stringent standards of the construction industry and yet satisfy the desire to build the project from 100 percent recyclables. Terri Musser worked diligently in researching and discovering virtually the entire product line that was used in the development of the project. A total of twenty-one separate products made from recyclables were used to build the 0.8-mile-long greenway.

Putting Recyclables to Work

The most readily available product for greenway construction was recycled plastic lumber that was used in the construction of bridges, benches, and signposts. In the early stages of the project, lumber from recycled plastic was not yet defined by "dimensional" standards, which made it a bit more difficult to work with in the building of greenway elements such as bridges and benches. Also, much of the plastic lumber supply available in the marketplace was of fairly low quality. Plastic lumber is also extremely heavy and elastic, and we had to adapt our designs for both these conditions.

Through extensive research, several companies were chosen that demonstrated a capability to deliver high quality plastic lumber and other products to the project. One by one, each product emerged in a similar manner, until every product used in the project originated from recycled waste and was backed by a performance guarantee. For example, signposts were made from postindustrial plastic diaper waste, educational and regulatory signs were made from recycled aluminum soda cans, and drainage pipes were made from recycled plastic grocery bags. Each product used was thoroughly researched and sanctioned for use.

The most significant recyclable product used in the project was bottom ash from Duke Energy (formerly Carolina Power and Light). Bottom ash, which is different from fly ash, is an inert granular material

The author explains how the trail bridge was constructed from recycled household plastic waste material. Source: Author.

derived from the burning of coal. Greenways Incorporated discovered that bottom ash would make a good substitute for aggregate base course (ABC)—the stone that is traditionally used as the foundation for surfaced trails, roadways, and highways. Working with Duke Energy, we submitted samples of bottom ash to the State of North Carolina in order to certify that it could be safely used in the environment. Our team was able to demonstrate that bottom ash would provide construction values similar to stone. As an added benefit to the development costs, replacing ABC with bottom ash didn't cost a penny. Duke Energy was mandated by law to dispose of the material in an approved landfill and paid a disposal fee to the state. The use of the bottom ash resulted in a cost savings equivalent to one-fourth of the total project value.

As with many publicly funded projects, the largest single obstacle that had to be overcome was funding of project development. We immediately recognized that if the Swift Creek Recycled Greenway was going to be successful, our public-private partnership would have to be expanded. With assistance from the town of Cary, we assembled a twenty-four-member partnership comprising nineteen private sector companies and five public sector agencies, including:

Duke Energy
C. C. Mangum, Inc.
Carolina Tank Lines
Gordon Bunn Landscape Contractor
McQueen Construction Company
Omer G. Ferrel and Son Grading Contractors
Sunshares Recycling
Amoco Fabrics and Fibers
Tim Helton Grading Company
Crumpler Plastic Pipe, Inc.
Earth Care Products, Inc.
BTW Industries, Inc.
ARW Polywood, Inc.
Reynolds Aluminum, Inc.
Tipper Tie
Innovative Plastic
Cary Public Works Department
Cary Planning Department
News and Observer Publishing
Edwards and Broughton
Wake County Solid Waste Management
Wake County Parks and Recreation
Wake County Field Services

The partnership enabled us to build the project for approximately one-half of traditional greenway development costs. All the project partners were honored to participate in building this first-of-a-kind greenway, and many devoted significant time, resources, and labor for which they were not compensated. Completing a successful project became the operating philosophy for the project instead of focusing on the traditional profit and loss associated with participation. The total public investment in the nearly one-mile trail was $99,240. Private sector donations exceeded $40,000 in value. Traditional construction costs at that time would have exceeded $200,000.

Beyond the two founding partners, three other partners, Duke Energy; C. C. Mangum, general construction contractor; and Sunshares Inc., a local recycling company, devoted significant time, energy, and creativity to the project. The partnership focused its energies on solving problems and developing quality-oriented trail development solutions. For example, we needed to find a way to mulch the landscape disturbed by trail construction. We shredded and chipped the woody material that was removed from the corridor opened for trail construction but found that there was not enough harvested material to properly cover all disturbed areas. It was suggested that we contact the Raleigh News and Observer, a local newspaper publisher, and ask them to donate shredded newspaper to complete our mulching. They happily joined our partnership.

Several partners published the results of their efforts in national trade journals and magazines. C. C. Mangum, the prime construction contractor, wrote a technical specification for the *Associated General Contractors of America* magazine describing the use of crumb rubber and roof shingle tabs in preparing recycled asphalt. Greenways Incorporated and the town were featured in *American City County* magazine for our leadership efforts. The project also received several awards, including North Carolina's 1992 Take Pride in America Award, and the United States Department of Transportation, Federal Highway Administration's 1995 Environmental Excellence Award.

Closing the Loop in Recycling

Educating the public about the benefits of greenways and recycling was the ultimate goal of the partnership. Our team came up with a slogan for the project: "Buy Recycled. It's Second Nature." We wanted to be certain that greenway users would leave the trail environment not just refreshed and renewed, but also more knowledgeable about the truths and facts about recycling. Sunshares, Inc., pooled its knowledge of recycling with Greenways Incorporated's talented graphic artists to develop an educational signage program for the trail. Topics range from the recyclables that were used to make a typical sign to the use of rubber tires in the preparation of the asphalt trail. Educational signs are displayed at strategic points throughout the greenway corridor.

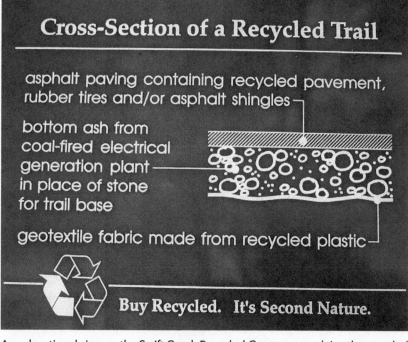

An educational sign on the Swift Creek Recycled Greenway explains the recycled materials used to build the asphalt trail. Photo by author.

The partnership also conducted numerous guided tours for small groups of citizens, business leaders, and elected officials. Education is the key to understanding the wealth of possibilities with greenways and recycling. The Swift Creek Recycled Greenway's educational program is one of the hallmarks of the project.

To participate in a community recycling program is not, in and of itself, all that is required to live sustainably. But it can be regarded as one of the more important initial actions. In order to better understand the world in which we live, and the essential interrelationships that affect our quality of life, each community and each citizen must understand the processes and systems that affect daily life. If we are to achieve a sustainable lifestyle, citizens must be made aware of the activities and choices that are essential to this new way of life. It is about changing the way we do business. Status quo will not lead us toward a sustainable future. Major changes in lifestyle are required, but it does not necessarily mean that these will result in enormous sacrifices to our economy or the quality of daily life. It does require a new approach to thinking about traditional practices and activities.

The Swift Creek Recycled Greenway is a small, yet significant effort that illustrates how individuals and a community can make a difference. The greenway satisfies the triple bottom line by being good for the environment, a positive influence on society, and an economically efficient development project. Recycling waste and thinking of trash as a resource for the future is principally a mindset, not a hardship. There are dissenting opinions regarding the effectiveness of recycling. There are those who claim that recycling of waste is not worth the effort or cost effective. From a professional and personal experience, I will side with McDonough and Braungart that we have to think and act differently and incorporate a closed loop mentality when it comes to resource and material use. To simply never think of any functional item within our home, school, or place of business as waste is a major step in the direction of living a sustainable life.

Assessment and Evaluation

During the past twenty-seven years the Swift Creek Recycled Greenway has performed well and satisfied the original vision and goals for the project. The quality of products used in project development has proven equal to or better than products manufactured from traditional construction practices. For example, the recycled plastic lumber that was used to build bridges, benches, and signposts has withstood weathering and aging better than traditional wood products. The greenway is located in a floodplain, situated in the Southeastern United States, and is subject to a wide range of temperatures, humidity, and exposure to sunlight. Despite being subjected to these environmental conditions, the project has held up well through the years.

It is typical for trails to be rebuilt or resurfaced after fifteen to twenty years. The Swift Creek Recycled Greenway trail was built using recycled asphalt milled from local streets and combined with crumb rubber from automobile tires and asphalt roof shingle tabs. The trail surface has remained resilient and functional. It certainly is no worse for wear than any other comparable asphalt greenway trail in the Triangle region of North Carolina.

The educational signs are worn and should be replaced but have held up well considering the environmental conditions. The trail furniture and other amenities are also weathered but have outperformed similar trail amenities of similar age constructed from wood.

The Swift Creek Recycled Greenway has stood the test of time. The project has demonstrated, both in the short term and long run, that using recyclables to build infrastructure for the future is a wise investment of time, energy, resources, and community effort.

Perhaps the more important and nagging question is why aren't more greenways constructed from recyclables? It is an appropriate question given the success of this project. I have several thoughts and conclusions that are worth sharing:

Too Little Demand and No Market

The United States has not achieved a critical tipping point of demand for recyclables to replace materials and goods manufactured from virgin materials. Until we begin to tip the scales in the marketplace, the use of recyclables will remain a somewhat token effort in the planning, design, manufacturing, and construction process. McDonough and Braungart emphasize the need for an overhaul of the design process in order to achieve this. They point to both Ford Motor Company and Nike as innovators in the field of reengineering to accommodate a design philosophy that employs cradle-to-cradle architecture and manufacturing. More and more quality information is being provided to the planning, landscape architecture, and engineering community with respect to building greenways from recyclables, but more demand is required on the part of both clients and designers in order to effect change.

Quality of Material

Many in the recycling field are quick to point out that given the low participation rate in recycling, it is difficult to control the quality of material in the reuse stream. There are two important factors or concerns in this regard. First, when it comes to the sorting of trash to remove waste that cannot be recycled, this process is generally inferior to the processes used to supply virgin material from extraction or industrial production. Contamination of the waste stream comes from the fact that consumers are not taking the time to educate themselves on local recycling procedures, and in some cases the governments or companies in charge of recycling are not providing consumers with such helpful information.[11] The procedures will be different for curbside recycling and drop site recycling.[12] Second, recyclables are commingled, and quality is not the first order of business when it comes to supplying, for example, plastics from the waste stream. With only 25 percent of all plastics being recycled, manufacturers that desire to utilize the waste stream to build new products are forced to use mixed-grade plastics or go through a second sorting process that drives up the cost of product development.

It Is Not Second Nature to Buy Recycled

To be clear, some products are easier to purchase than others when it comes to using manufactured items made from recyclables. For example, in our greenway projects, it is common for us to use recycled plastic wood for benches, trash receptacles, drainage pipe, and signs. However, recyclers and manufacturers have a long way to go to provide the full range of products that consumers want and need to build 100 percent of their projects from recyclables. From a sales point of view, very little is being done to distinguish and promote products made from recyclables.

It Is Easier to Extract Than to Recycle

It is far easier and more of an entrenched institutional practice to continue mining, harvesting, and extracting raw materials from the Earth than to recycle trash into materials and building supplies for manufactured products. In order to complete the transition, the economics associated with purchasing and using products manufactured from virgin materials versus products produced from recycled waste will have to change. Perhaps the most important trigger for this change will be the availability and pricing of petroleum. As long as the price of a barrel of oil remains affordable, it is more cost effective to continue the extraction process. As the price of oil rises, institutional, social, and economic change will take place. When the pricing swings in favor of products made from recyclables, we will see more projects constructed from recyclables.

Further, China, the most significant recycler in the world, has instituted restrictions on what they are willing to accept due to the careless commingling and poor quality of waste being delivered by American and European nations. Recyclables are an important export of the United States, and much of the nation's recyclables are shipped to China for repurposing. The Chinese have instituted a new policy called "Green Fence," which places quality restrictions on the type of recyclables that can be exported.[13]

Dare to Be Different

Finally, it may simply come down to Americans striving to forcibly change habits and begin the process of building more projects using recyclables. As a company, we make an effort to incorporate recyclables in all of our projects. Right now, that is easier to accomplish with products made from recycled paper, plastic, and some metals. With an increased interest in sustainable planning and design, the desire to achieve market success is favoring the use of more recyclables. Americans must strive to innovate whenever and wherever possible to effect change. The Swift Creek Recycled Greenway proved that it can be accomplished, and we must now replicate the process repeatedly.

The Urgency at Hand

When it comes to managing human waste streams, it is fair to conclude that we must all begin to act with a sense of purpose and urgency to reduce the amount of trash we generate and change the culture around waste management. Earth's natural systems are being overwhelmed by the amount of waste humans generate. For example, large marine animals are continuing to wash ashore killed by consuming huge amounts of plastic that are floating in our oceans. Our lack of desire to treat waste as a resource for repurposing and the negligence by which we discard our waste across the landscape, into our streams and oceans is compromising many forms of life across the planet.

Humans' inability to manage waste is not limited to the confines of our planet, as we now have enormous amounts of discarded material and debris circling our planet, remnants of the space age, which is barely sixty years old. According to the National Aeronautics and Space Agency, approximately 500,000 pieces of human-generated trash and debris larger than a marble, labeled as space junk, circles our planet every day and makes space travel hazardous. Much of this debris is traveling at speeds of around 17,500 miles per hour and can easily damage spacecraft and satellites. Space junk can disable important human infrastructure that powers our telecommunications and America's national defense systems. The point in raising this issue is to define the sense of urgency needed to manage all human-generated waste streams. An entirely new approach and nothing less than a cultural change is needed to correct the problems associated with repurposing waste.

The story of the Swift Creek Recycled Greenway is one rather small effort. Other more significant efforts are already under way. In the Netherlands, the Dutch are experimenting with the construction of new bike paths made entirely of recycled plastic. The Dutch government understands that recycling waste is part of a new "circular economy" in which there is no waste, but only a continual repurpose of human-made products. On the flip side, it is worth noting that in 2018, China, the largest importer and manufacturer of products made from recycled trash, began turning away waste shipments from Western nations. The Chinese are overwhelmed by the amount of waste already collected. It is a signal to Western nations that it is now their turn to begin investing in the technology, machinery, and processes that are required to transform their waste streams into repurposed products for consumer use.

4

Something Grand

Grand Canyon Greenway,
Grand Canyon National Park, Arizona

Peter Axelson and the author lead a tour of the newly opened Phase 1 of the Grand Canyon Greenway along the South Rim of the National Park. Source: U.S. Department of the Interior National Park Service.

ON A BRIGHT, CRISP, AND WINDY WINTER MORNING in January 1997, I made my way to the edge of Grand Canyon's South Rim to witness the first rays of sunlight illuminating more than six million years of geologic history on the North Rim walls. Snow and ice covered the ground, and as we adjusted to the bitterly cold dry air my colleagues Jeff Olson, Bob Searns, Dan Burden, Andy Clarke, Charlie Gandy, Ben Pugh, Mark Fenton, Peter Axelson, and I listened to the wind and soaked in the warm rays of winter sunshine. Not a word was spoken. We were overwhelmed and humbled by the raw beauty, stunning landscape, and majestic views. The morning interlude provided us with time to reflect and grasp the enormity of the task in front of our group. We were simultaneously excited, curious, skeptical, confident, but mostly ready for the challenge. Our all-volunteer design team was venturing into rarely trodden territory, and we were using Grand Canyon National Park as our laboratory.

As Teddy Roosevelt stated in 1903: "In the Grand Canyon, Arizona has a natural wonder which, so far as I know, is in kind absolutely unparalleled throughout the rest of the world. I want to ask you to do one thing in connection with it in your own interest and in the interest of the country—to keep this great wonder of nature as it now is. Leave it as it is. You cannot improve on it. You can only mar it. What you can do is keep it for your children, your children's children, and for all who come after you, as one of the great sights which every American should see."[1]

President Roosevelt's quote is as relevant today as when those words were first spoken at the Grand Canyon in 1903. Despite their efforts to create points of access from which visitors could view this natural wonder, National Park Service planners were faced with a problem common in modern national parks—a crush of visitors crowding into undersized overlooks to get a glimpse of the majestic view. It is a cliché commonly referred to as "a Kodak moment." The National Park Service staff shared with us that many of the 5.5 million visitors to the South Rim typically spend less than ten minutes viewing and experiencing the Grand Canyon. Many visitors stand with their backs to the gaping landscape, they take a few selfie photos, chat with friends or relatives about the awesome experience, and then load back into the cars or busses they arrived in and embark on a long return journey. Our team wanted to help the National Park Service change the visitor experience by building greenway trails along the rim of the Canyon that would extend the stay for more than a few minutes and enable visitors to more fully experience the awesome beauty of Grand Canyon National Park.

America's National Parks

America's national park system has several points of origin. One account has the concept originating with George Catlin in 1832,[2] an artist best remembered for his paintings depicting American Indians. Catlin was concerned with the destruction of landscapes and Native American settlements in the western United States and proposed establishing parks as a way to protect these valued resources: "by some great protecting policy of government . . . in a magnificent park . . . a nation's park, containing man and beast, in all the wild[ness] and freshness of their nature's beauty!" Other accounts define the awareness and contributions to the idea of protecting American landscapes to Thomas Jefferson's 1785 writings about the natural beauty of Virginia, James Fenimore Cooper's (author of *The Last of the Mohicans*) essay on the scenery of the United States and Europe, and landscape architect Frederick Law Olmsted's advocation for the protection of Niagara Falls.[3]

In 1864, California took steps to protect the Yosemite Valley. Frederick Law Olmsted's vision for Yosemite was instrumental in convincing California's legislature that protection of Yosemite was in the interests of the state. In a report to the legislature, Olmsted stated that it is the "political duty" of a republican government to set aside "great public grounds for the free enjoyment of the people." This led to Congress and Abraham Lincoln agreeing to legislation that protected Yosemite "for public use, resort and recreation . . . inalienable for all time."[4] Eight years after Olmsted's assessment of the Yosemite Valley, another scenic valley, Yellowstone, located in the northwest corner of Wyoming, became America's first National Park in 1872.

Under the leadership of Frederick Law Olmsted Jr., the 64th United States Congress ratified the 1916 Organic Act that established a unified system of national parks and a professional bureau to manage them, which came to be known as the National Park Service (NPS). Olmsted Jr. was instrumental in bringing his father's vision in Yosemite to reality in the creation of a national park system, professionally managed and protected for all time. Upon its establishment, NPS assumed management of the lands and waters that comprised Grand Canyon National Park. First afforded Federal protection in 1893 as a Forest Reserve, on February 26, 1919, Grand Canyon National Park became the fifteenth federally recognized park, when it was added to the system by President Woodrow Wilson.

There is arguably no greater legacy for the profession of landscape architecture than the development of America's national park system. Landscape architects were the visionaries who argued for the preservation and protection of America's wild lands and waters. Landscape architects crafted national policy and legislation that created the capitalized national park system. Landscape architects have been leaders in preparing the master plans, general management plans, and construction documents that guided the construction of facilities that enable hundreds of millions of people to experience America's national parks each year. As a licensed landscape architect since 1986, I felt considerable pride in the heritage of my profession as I joined a team of volunteers to work on a greenway development strategy that we all hoped would leave a lasting legacy at Grand Canyon National Park.

Conflicts in Visitor Experience

In the early 1990s, NPS officials deemed the visitor experience at Grand Canyon National Park to be unsatisfying, disappointing, and even potentially hazardous. Vehicle traffic overwhelmed park roads, overlooks were congested, vehicle noise damaged the solitude of the experience, and poor air quality diminished the visual experience. Occasionally fistfights would break out among visitors coveting one of the few available parking spots. In some cases, weapons were brandished over the right to a parking space. Most important, movement around and within the park was a significant challenge for visitors. The park transportation infrastructure was built to service automobile travel. To experience the park as a pedestrian was difficult at best, and at times dangerous. Constructed paths for pedestrian travel were few and far between. Most pedestrian routes were random worn footpaths, resulting in a braided network of trails scattered throughout the landscape. Bicycling within the park was strictly prohibited.

The thinner air at the 7,000-foot elevation along the South Rim affects many who visit the park. People with disabilities are challenged to enjoy the awesome beauty of the park as their path of travel, in certain locations, can be even more difficult. Much of the park was constructed before accessibility became a national imperative. It was these experiences and those moments of interaction with the park ecosystem that we the volunteer designers wanted to change, and we proposed to use a greenway strategy to alter the visitor experience.

Motorists look for parking along the South Rim of Grand Canyon National Park. Photo by author.

A World Heritage Site

Grand Canyon National Park is one of the most popular of the fifty-eight national parks in the United States. It was designated a UNESCO World Heritage site in 1979, the only landscape in North America so recognized at that time.[5] Much of the modern park infrastructure was planned, designed, and constructed first by the Atchison, Topeka and Santa Fe Railroad at the turn of the century and then by the federal government in the 1930s, '40s and '50s. Railroads played a key role in the development of our national parks, as they understood the appetite of Americans for adventure travel and tourism, and capitalized on the opportunity to open western wilderness lands to service these interests. In 1905, the railroad opened a luxury lodge, El Tovar, on the edge of the South Rim. Designed by the railroad's architect Charles Whittlesey, El Tovar was designed as a hybrid Swiss chalet and Norwegian Villa to appeal to predominantly European values of the day. For many years, El Tovar was regarded as the most elegant hotel west of the Mississippi River.

Cover of the Grand Canyon National Park General Management Plan. Source: U.S. Department of the Interior National Park Service.

In 1966, in conjunction with the fiftieth anniversary of the national park system, Mission 66, led by landscape architect and NPS director Conrad Wirth, was launched by the federal government to make wholesale improvements to infrastructure, employee housing, and visitor facilities within the national parks. Despite incremental investments through the years, the popularity of the Grand Canyon has continued to overwhelm the infrastructure of roads, lodging, and viewing areas.

In 1995, the Park Service completed a General Management Plan (GMP) for the Grand Canyon National Park. According to the National Park Service, General Management Plans "support the preservation of park resources, collaboration with partners, and provision for visitor enjoyment and recreational opportunities. These plans provide the basic guidance for how parks will carry out statutory responsibilities for protection of park resources unimpaired for future generations while providing for appropriate visitor use and enjoyment."[6] These plans are typically prepared by the park superintendent, with input from landscape architects, ecologists, natural and cultural resource specialists, and others who are involved with park concessions, interpretation, and park management.

In addition to leadership from Grand Canyon park superintendent's office, portions of the GMP were prepared by Shapins Associates, a landscape architecture firm based in Boulder, Colorado. Among the many recommendations of the GMP for Grand Canyon National Park was a set of major infrastructure initiatives for the North and South Rims that included an extensive network of bicycle and pedestrian trails, which we used to form the basis of the Grand Canyon Greenway.

A Wedding and a Vision

The ringleader of our volunteer group was Jeff Olson, an architect and planner whom I met briefly in 1994 when he was the New York Department of Transportation bicycle and pedestrian program manager. After our work at the Grand Canyon, Jeff, a graduate of both Rensselaer Polytechnic Institute and the State University of New York—Empire State College, went on to serve as director of Millennium Trails, an award-winning initiative of the Clinton White House that was to create a national network of trails as part of America's legacy for the year 2000.

Jeff's wife Margo was at one time a park ranger at Grand Canyon National Park. Margo maintained friendships with other park rangers, and in the summer of 1990 Jeff and Margo attended the wedding of one of those friends. Jeff discovered during the wedding weekend that riding bicycles within Grand Canyon National Park was prohibited and the same rules applied to many other national parks. He thought this was ludicrous. Why would the federal government prohibit the use of bicycles within our national parks? He was determined to change this policy. Also in attendance at the wedding was Brad Traver, park planner at Grand Canyon National Park, who was leading efforts to prepare the new General Management Plan to guide future growth of the park. The concept for a network of bicycle and pedestrian trails at Grand Canyon had been under consideration by National Park Service staff. When Jeff received a copy of the 1995 GMP, he viewed a graphic that in his opinion depicted an improved network of bicycle and pedestrian trails that he interpreted to be a future greenway system.

Seizing the opportunity, Jeff wrote a letter to park superintendent Robert Arnberger voicing his support for the future greenway network and offering his services, as a volunteer, to transform the concept of a greenway into reality. Jeff later spoke by phone with Brad Traver, who asked him: "Who are you? Are you serious about volunteering? When can we meet to discuss this in more detail?"[7] Jeff was already working with world-renowned walking and bicycling guru Dan Burden on a project to make the Las Vegas Strip more friendly to pedestrians and cyclists. They agreed to meet Traver at a hotel in Las Vegas to discuss Jeff's vision for a proposed Grand Canyon Greenway. At that meeting, Brad introduced Jeff to Robert Koons, executive director of the Grand Canyon Association. The association was founded in 1932 by naturalist Edwin D. McKee and as a nonprofit raised funds for specific park projects and initiatives to enhance the visitor experience. Jeff asked if the association would serve as an official sponsor for a greenway master plan effort. Koons agreed, and plans for a Grand Canyon Greenway Summit began to take shape.

An Invitation to Volunteer

In September 1996, at the ProBike/ProWalk conference in Portland, Maine, Jeff Olson asked conference host Bill Wilkinson, executive director of the Bicycle Federation of America, to make a simple announcement to the assembled participants. It went something like this: "Anyone who wants to volunteer to work on a greenway master plan for the Grand Canyon National Park meet in room XXX after the plenary session has concluded."

Jeff is very persuasive. His vision and energy are infectious. My thought at the time was "who was going to say no" to an opportunity to work on a master plan for the Grand Canyon Greenway? I was one of the first to arrive at the meeting, and in the nearly empty room I could see Jeff Olson and Bob Koons seated at a table. As I recall, Peter Axelson, Charlie Gandy, Andy Clarke, and Dan Burden also attended. Jeff asked who else should be a participant in the greenway summit, and I recommended that my friend and colleague Bob Searns be invited to join the group. What I vividly recall about that meeting was how *few* people attended. I thought this was the opportunity of a lifetime. I could tell that there was some disappointment in the turnout. What we lacked in numbers, our small group more than made up with enthusiasm. The only detail to be finalized was the date for the volunteer greenway planning summit. Bob Koons spoke briefly about the financial support that would be provided to the volunteers by the association, which covered the cost of travel, lodging, and meals. Jeff shared that he also had the support of Superintendent Arnberger. There was little doubt that this was an out-of-the-box solution—volunteers developing a greenway strategy for a national park. I got the impression that both the association and National Park Service had toes in the water but were still somewhat skeptical of the undertaking. We, the volunteer team, despite impressive resumés and prior accomplishment, had much to prove.

January at the South Rim

Jeff immediately went to work securing a date and the financial backing to support the work of ten volunteers. On January 24, 1997, we began arriving at Phoenix SkyHarbor airport and made the three-hour drive to the South Rim of Grand Canyon National Park. I volunteered to pick up art supplies in Flagstaff, Arizona, so that we would have the resources necessary to illustrate our ideas and recommendations. This was before the age of modern computer software, and we would rely primarily on hand drawings and renderings to communicate our ideas. Computers at that time were mostly used for word processing and were in the early stages of becoming more robust in digital presentation, desktop publishing, and computer-aided drafting.

It was absolutely the coldest time of year to be at the South Rim, and the park was empty for the most part, except for a few visitors and permanent staff. The Park Service provided access and use of facilities at the Horace M. Albright Training Center. The center was equipped with the latest in technology, which greatly aided our work and the ability to present our recommendations using digital media.

A Great Group

It is important that I introduce to you the incredible team of people I worked with whose imagination, skills, expertise, and hard work brought the idea of the Grand Canyon Greenway to life. Jeff Olson was the

The design team is photographed on the South Rim of Grand Canyon National Park. Source: Author.

instigator, but he managed to attract an all-star team of professionals who contributed a significant amount of volunteer time to meet the challenges of the project.

Peter Axelson is a mechanical engineer with degrees from Stanford University and is one of the world's foremost experts on barrier-free, fully accessible design. Peter was paralyzed from the waist down during a climbing accident while he was a cadet in the United States Air Force Academy in Colorado Springs and took on the challenge of his disability to become an inventor, designer, advocate, and world-class athlete. Peter founded Beneficial Design, Inc., in 1981 as a firm that specializes in the design, development, and testing of adaptive equipment for people with disabilities and was a critically important member of the Grand Canyon Greenway team. At the time of the greenway summit, Peter was on the board of directors for American Trails, the Recreation Access Advisory Committee to the Access Board (U.S. Architectural and Transportation Barriers Compliance Board) and the chairman of the ANSI/RESNA Wheelchair Standards Committee. By virtue of Peter's leadership and design recommendations, the Grand Canyon National Park was awarded the 2004 Outdoor Accessibility Award from the Department of the Interior in recognition of the manner in which the Grand Canyon Greenway improved access for all people.

Dan Burden was an incredibly accomplished professional when he joined the greenway volunteers in 1997 and easily the most well known. He had worked as a photographer for the National Geographic Society, was the Florida State Bicycle and Pedestrian Coordinator and had embarked on a career to make America, and indeed the entire world, more bicycle and pedestrian friendly. In 2001, *Time* magazine named Dan "one of the six most important civic innovators in the world." A 2009 user poll by Planetzien named Dan as one of the Top 100 Urban Thinkers of all time. In the 1970s, Dan cofounded Bikecentennial with his wife Lys and led a bicycling expedition from Alaska to Argentina. He and Lys also worked with ninety governmental agencies to develop the longest recreation trail in the world: the 4,300-mile-long TransAmerica Bicycle Trail. In 1977, Dan founded the Bicycle Federation of America and served as its director for its first two years of operation. Beginning in 1980, he served as the country's first statewide bicycle and pedestrian coordinator, which soon became a model for other statewide programs throughout the nation. Prior to his arrival at the Grand Canyon, Dan worked as a bicycle consultant in China for the United Nations in 1994 and in 1996 formed a new company called "Walkable Communities," which promoted the benefits of bicycling and walking and introduced the concept of "road diets"—slimming down bloated auto-oriented streets to create safe spaces for cyclists and pedestrians.

In 1997, Andy Clarke was vice president of Trail Development for the Rails-to-Trails Conservancy (RTC), a national nonprofit organization with 65,000 members at the time of our South Rim gathering. Prior to his work at RTC, Andy was the deputy director of the Bicycle Federation of America where he managed many of the BFA bicycle and pedestrian planning and research contracts. He authored several of the National Bicycling and Walking Study case studies for the Federal Highway Administration. Andy later served twelve years as president of the League of American Bicyclists. He joined the league in 2003 to establish and implement the Bicycle Friendly Community program, taking over as the CEO in 2004. A native of England, Clarke has also worked for Friends of the Earth (UK) and the European Cyclists Federation. During a twenty-year period Andy wrote groundbreaking case studies and reports for the

Federal Highway Administration and was involved in nearly every major initiative and development regarding bicycling and walking policy and planning, including the formation of the National Complete Streets Coalition, the Safe Routes to School National Partnership, multiple federal transportation funding bills, and Vision Zero.

Prior to joining our volunteer group in 1997, Charlie Gandy had been named one of the "30 Most Influential People in the Bike Industry" by *Bicycle Dealer Showcase* magazine. Charlie was director of advocacy programs for the Bicycle Federation of America, where he organized and launched citizen-based advocacy groups for walking and cycling in thirty states and metropolitan areas. Gandy coached and trained advocates in all fifty states. In 1998, Charlie was named "America's #1 Bike Advocate" by *Velo Business* magazine. Charlie served as a member of the Texas House of Representatives for Dallas from 1983 to 1985. As a member of the House he passed several bills improving public safety and was a leader of Texas Education Reform. Gandy later founded and served as the first executive director of the Texas Bicycle Coalition from 1990 to 1994. In 2009, Gandy became the mobility coordinator for Long Beach California's Bike Long Beach program, which is where he lives and works today.

Betty Drake was our local connection to Arizona. At the time of the summit, Betty was chair of the Arizona State Committee on Trails, and she was a faculty associate at Arizona State University, College of Architecture and Environmental Design. Betty is a pioneer in bicycle and pedestrian facility development with twenty-five years of private practice. She worked with Clare Cooper Marcus on landmark studies of pedestrian activities in San Francisco parks and neighborhoods and with Donald Appleyard on the Street Livability Study. Betty completed work in more than fifty U.S. cities, including the 1974 Tempe Bikeway Plan, which received the Historic Arizona Planning Landmark Award from the American Planning Association and the Eastern Pima County Trails Master Plan.

Mark Fenton was the exceptional athlete of our group, an Olympic racewalker and member of the United States Olympic team from 1986 to 1991. He competed in the 1984 and 1988 Olympic Trials in the 50-kilometer racewalk. When he joined our group he was employed by *Walking* magazine as a writer and editor-at-large and was hosting a series on PBS television titled *America's Walking*. Mark studied biomechanics at the Massachusetts Institute of Technology and was a researcher at the Olympic Training Center's Sports Science Laboratory in Colorado Springs. Years later, Mark served with me on the East Coast Greenway Alliance Board of Trustees. Today, Mark is a national public health, planning, and transportation consultant and an adjunct professor at Tuft's University's Friedman School of Nutrition Science and Policy. Mark is one of the most sought-after speakers on health and wellness and has authored numerous books, including the best-selling *Complete Guide to Walking for Health, Weight Loss and Fitness*.

Ben Pugh was the kind of civil engineer who complemented our volunteer team. Ben graduated from Oregon State University in 1958, with a BS degree in civil engineering. He was a registered civil engineer, a licensed land surveyor, and a registered traffic engineer. He served ten years as a highway engineer with the Federal Highway Administration, specializing in preliminary engineering and construction of roads and bridges located in virgin areas of national parks and other federal lands. He had more than twenty years of experience as a civil engineer with Sacramento County, which developed 300+ miles of on-road and off-road bikeways, including the American River Greenway. He was also involved with the Sacramento Northern Railroad Bike Trail and the Sacramento City/County Bikeway Master Plan, which details the implementation of 1,124 miles of on-road and 205 miles of off-road bikeways.

Bob Searns is one of America's most accomplished greenway planners and designers, having worked on projects across America and around the world. Bob coauthored with me two award-winning publications: *Greenways: A Guide to Planning, Design, and Development* and *Trails for the Twenty-First Century*. Bob's early career in greenway planning and design began when he served as project director of Denver's Platte River Greenway. He later developed the award-winning Mary Carter Greenway in Littleton, Colorado. Bob has been an instructor and advisor for the Urban Land Institute and American Planning Association Mayors conferences, National Park Service, the National Recreation and Park Association, American Rivers, and the National Rails to Trails Conservancy. After our work at the Grand Canyon, he served as chair of the board of trustees of American Trails, Inc.—the nation's leading recreational trails advocacy organization—and was a delegate to President Barack Obama's White House Summit on America's Great Outdoors in April 2010.

In addition to the volunteers, we were graciously welcomed and supported by superintendent Rob Arnberger, park planner Brad Traver and a landscape architect from the NPS Denver Service Center, Bob Pilk. There were numerous other Park Service staff who played instrumental roles in the success of the project. To this day, this was the greatest collection of people that I ever had the chance to work with on any project throughout my career.

Entering the Digital Age

Another critically important detail to describe was our ability to use digital technology as a platform for generating finished work. While it is commonplace today and so much of our world operates on digital infrastructure, that was not the case in 1997 as much of the original technology was just making its way into the marketplace. As the owner of a small company, I knew that the most efficient way to grow my company was to invest in technology. As a 1996 Christmas gift to myself, I purchased my very first digital camera and brought it along to test it out at the Grand Canyon. As I recall, it was manufactured by Kodak and was a fixed lens camera that took pictures in two modes: 640 × 480 pixels per inch and 320 × 240 pixels per inch, which by today's standard is very low resolution. Most smart phones have significantly higher resolution today. There was no external storage card for the camera, and when switched to its high resolution mode the camera could store a limited number of images. The camera had a small LCD screen on the back that one could use to view and delete images. I could connect the camera to my Apple Powerbook Duo laptop computer, using the computer's serial port, to import images for use in documents and presentations.

Digital cameras in 1997 were new and not widely used in the design industry as they had severe limitations in terms of image quality, storage, and functionality. Serious professionals interested in high quality images used film cameras. A digital camera was more a novelty item than a graphic arts tool. However, I realized that our team was going to be sequestered at the South Rim of the Grand Canyon, in January, with limited access to services that landscape architects would typically count on, such as photo processing labs, commercial printers, and document production companies. I thought that a digital camera might come in handy, so I packed it for the trip. Little did I realize at the time the impact that this one seemingly insignificant device would have on the work of the volunteer team.

I will always remember Dan Burden's first reaction to seeing the digital camera at work. I was having fun, taking photos of everyone's work and importing the images for use in design and presentation software. Dan thought that this was amazing—mind boggling actually—as no one on the team had seen one of these contraptions at work. It was quite the compliment coming from Dan, whose spectacular photography had been published in the *National Geographic Society*. He interrupted my work and asked, "Chuck, this is really exciting, and I have one question. Can you tell me what kind of film the camera uses?" Our entire team erupted in laughter as I explained to Dan that the camera did not use film and that the images were digital, a series of ones and zeros that provided the code for each pixel. It was an incredibly difficult concept for Dan to grasp, as someone who had accumulated tens of thousands of slides throughout his professional career.

For the entire team, the digital camera became a magical device. We were able to generate hand sketches of our recommendations, which I would then photograph and incorporate into our presentations to the National Park Service staff. For the designers this was a first as the digital camera provided us the luxury of additional time and the ability to make last-minute adjustments to our presentation. We loved the new digital medium and the way it produced high quality work. For me, it was a turning point in my professional career as I understood the power of working solely within a digital environment and began to invest more heavily in the technology that enabled my firm to generate all our work in a digital format.

A Vision and a Plan

Jeff Olson divided the volunteer team into two teams: design and funding. Design team members included Brad Traver, Dan Burden, Ben Pugh, Peter Axelson, Bob Searns, Betty Drake, Bob Pilk, and me. The funding team included Jeff Olson, Andy Clarke, Charlie Gandy, Mark Fenton, and Robert Koons. I was asked to lead the design team, and Jeff led the funding team. The design team was tasked with exploring the landscape architectural, civil engineering, transportation engineering, and implementation strategies associated with the development of a seventy-two-mile system of trails along the South and North Rims of the Grand Canyon. The funding team was asked to prepare construction cost estimates and propose ways to fund greenway development.

Our design program for building the greenway system was intentionally simple, with the goal of using indigenous materials and a minimalist approach—less is more. Our approach was to provide a seamless, fully accessible pathway system along the edge of the canyon rim. It had to be safe for all people to use and at the same time not an intrusive element of the landscape. The trail tread had to be durable but also blend with the surrounding high desert landscape. We needed to develop the greenway so that it connected visitors to other features and destinations throughout the park. Our specific design solutions and recommendations included the following improvements:

An interpretive welcome center that would offer visitors a clear understanding of how best to experience the Grand Canyon on foot and bicycle.

An improved pedestrian trail network along the South and North Rims.

An expanded bicycle and pedestrian trail network that would offer alternative transportation throughout the park.

A facility that would make available bicycles, strollers, and wheelchairs for visitors, at a reasonable cost.

A signage and wayfinding system that included enhanced interpretive signs, information kiosks, and other signs to guide visitor exploration.

A center for improved transit throughout the park.

As I reflect, some twenty years later, on the above work program, it is gratifying to know that all of it has been accomplished. The visitation experience at Grand Canyon is vastly improved as a result of this completed work program. Over the course of two workshops, one in January 1997 and the second in July 1997, the volunteer team, Park Service staff, and Grand Canyon National Park Foundation (GCNPF) examined three landscapes in greater detail and specificity: the South Rim, the Village area of the South Rim, and the North Rim.

South Rim Greenway

In January 1997, the volunteer design team surveyed the existing land-scape along the South Rim and found it disturbingly devoid of pedestrian and bicycle transportation accommodation. We discovered that the best way to transport yourself across the South Rim was by car or on one of the Park Service buses. If you wanted to walk to a destination, you either had to travel along one of the braided foot trails that had been estab-lished by visitors or walk along the congested and overburdened roads. The problem with the foot trails in 1997 was you had no idea where they would lead you. As mentioned earlier, bicycling was forbidden.

The team sketched a system of bicycle and pedestrian trails that served different functions and offered an opportunity to organize and guide pe-destrian and bicycle travel throughout the park. We also focused on ways to better connect with the existing park transit system.

We crafted a set of guiding design principles for the South Rim Green-way that included:

Be compatible with natural setting and minimize environmental intrusion.

Provide a continuous shared-use trail from South Kaibab trailhead to Hermits Rest.

Provide for short distance trips throughout the South Rim to meet the needs of various user groups.

Make certain that trails are wide enough to accommodate all users, including visitors with disabilities.

Offer trail users a sense of intimacy with the canyon and surround-ing high desert landscape.

Complete trail design to provide users with variety, rhythm, and syncopation that matched the surrounding landscape.

Minimize interface with roads and automobile traffic.

Tell the unique Grand Canyon National Park story through design and with interpretive signage.

Our team, with the help of Bob Pilk, sketched out a plan for a robust visitors center, which we felt should be located south of the popular Mather Point overlook. This visitors center was eventually built and today is known as Grand Canyon Visitor Center. It provides ample parking for all types of vehicles, an orientation plaza, theater, bookstore, and shuttle stop where visitors can board buses for tours of the park.

Village Greenway

Grand Canyon Village is the historic core of the South Rim and a National Historic District. Originally built by the Santa Fe Railroad Company, the Park Service was making efforts in 1997 to invest in a series of improvements that would enable the village to absorb the dramatic increase in visitation. The village is home to the historic El Tovar hotel and restaurant, the train depot and Kolb Studio, one of the most historic structures in the park perched on the South Rim canyon wall. It was the photographic studio of Emory and Ellsworth Kolb from 1904 to 1976 and, after the death of Emory in 1976, was transformed into a bookstore, gift shop, and visitor center.

As our team surveyed the village landscape, we again found it devoid of pedestrian pathways and without accommodation for bicycle travel. Further, there were no connecting paths in and out of the village that would encourage visitors to walk or bike to other South Rim destinations. Our plan for the Village Greenway system consisted of an internal trail and rim trails that would help direct visitor travel and hopefully relieve confusion and congestion.

North Rim Greenway

In my opinion, the North Rim of the Grand Canyon is one of the most beautiful landscapes on planet Earth. Due to the success of our efforts at the South Rim, Superintendent Arnberger and the foundation eagerly invited our volunteer design team to visit the North Rim in July 1997 to work on greenway solutions for the North Rim peninsula. We scheduled our trip over the July 4th holiday. For every team member it was a magical experience. We came up with some forward-thinking recommendations that were put into service right away by the NPS.

Grand Canyon Greenway Development Plan

Grand Canyon National Park

Grand Canyon Greenway Summit
January 24-27, 1997
July 4-7, 1997

Cover of the Grand Canyon Greenway Development Plan. Source: Author.

Most of our work focused on connecting the North Rim's Grand Canyon Lodge with the North Kaibab Trailhead. An existing worn dirt path called the "bridle trail" served as a route of travel. We explored other routes as well and eventually proposed more than twenty-five miles of trails that extended along the Walhalla Plateau all the way to Cape Royal.

Presentations to the Superintendent

The January workshop time evaporated quickly. Crunch time was upon our group as we rapidly pulled together drawings and other graphic illustrations, recommendations, and funding scenarios for the Grand Canyon Greenway. We gathered at the Albright Training Center the evening of January 26 for one final late-night push.

The next morning, a bit bleary-eyed but very excited for the opportunity, our team made a professional presentation to the superintendent, foundation president, and invited Park Service staff. We didn't have time to rehearse the presentation, and it came off without a hitch. Superintendent Arnberger in particular was very moved by the work generated and the recommendations of the team of volunteers. As I recall, his observations and comments included the following: "As some of you know, I was born here at Grand Canyon National Park, I have been in the Park Service my entire professional career, and I have to be honest and tell you I have never viewed a more professional and visionary presentation. I am amazed at the work that you have generated in such a short amount of time, and furthermore I am excited about your ideas and recommendations. Thank you. You are to be commended for your efforts." It was an emotional statement, and it caught many of us by surprise. We had worked hard for several days, did not sleep much, and wanted very much to offer our best thoughts and recommendations to the Park Service. According to Superintendent Arnberger, we hit the mark. Margaret Mead's quote best summarizes the work and contributions of the volunteer Grand Canyon Greenway team: "Never doubt that a small group of thoughtful, committed citizens can change the world; indeed, it's the only thing that ever has."

Final Plan

After our July excursion to the North Rim was completed, it became apparent to everyone that we needed a report that would summarize the work of the team. I volunteered to produce one using desktop publishing software. I collected the supporting graphics, many of which I had photographed using my digital camera, and generated a summary report that captured the most important recommendations of the January and July workshops. Having a final report in hand was not only a good summary of the work; it provided legitimacy to the work and gave both Superintendent Arnberger and GCNPF President Koons a quick reference containing all the recommendations.

A Letter to President Clinton

Jeff Olson came up with an idea that the volunteers should pen a letter to President Bill Clinton, asking him for the financial support required to build the Grand Canyon Greenway. The letter was cleverly disguised as an invitation to a ribbon cutting and dedication event to be held on July 4, 2000. The entire team signed it with an RSVP at the bottom titled "The

Grand Canyon Fund–Greenway Team." Interestingly, the project was funded within a year of that letter being written as both Jeff and Andy Clarke skillfully guided a U.S. Department of Transportation Intermodal Surface Transportation Act of 1991 (ISTEA) application through the process and the Greenway was awarded federal transportation funding. Ironically, a few years later, Jeff Olson became the director of the National Millennium Trails program within the Clinton White House and in this capacity was able to make good on the commitment to an April 1999 dedication ceremony featuring First Lady Hillary Clinton.

Building the Greenway

With the success of the greenway master plan for both the South and North Rims of the park, the attention of our volunteer team, the National Park Service, and the foundation immediately turned to transforming vision into reality. My company, Greenways Incorporated, entered into a contract with the Grand Canyon National Park Foundation to complete design and construction documents for the first phase of the greenway. This contract included a proof of concept submittal, which transformed our master plan into a set of detailed design guidelines that could be used by any contractor who worked on greenway trail construction.

The Grand Canyon National Park Foundation took on the task of soliciting initial funding for greenway development, which was accomplished through a public-private partnership. Much of the funding was secured by Deborah Tuck, who in 2000 became the executive director of the Grand Canyon National Park Foundation after Bob Koons retired. Our volunteer team continued working closely with Deborah to identify and secure funding. At times, team members helped author grant applications, and we provided Deborah with materials necessary to solicit funds from private-sector sources. A partial list of some of the public and private funding for the project is defined below. The purpose in sharing this information is to illustrate the mix of funds that was pursued and secured for greenway development.

$40,000 from American Airlines through the National Park Foundation

$479,600 from the Arizona Department of Transportation (1998)

$869,800 from the U.S. Department of Transportation, Federal
   Highway Administration (1998)
$499,900 from the Arizona Department of Transportation (1999)
$1 million from the Nina Mason Pulliam Charitable Trust through
   the Grand Canyon Park Foundation (2001)
$25,000 from the Dr. Scholl Foundation
$29,400 from the Richard Haiman National Park Foundation

An Accessible Path of Travel

I worked closely with Peter Axelson from 1997 to 2007 and learned a great
deal from him about the needs and interests of all trail users, regardless of
their physical mobility or other impairments. Peter believes all individuals should have access to the physical, intellectual, and spiritual aspects of
life. He bristles at the notion that accessible pathway design means hard-
surfaced trails everywhere with running slopes less than 5-percent and
cross-slopes less than 2-percent. Peter stressed that persons with varying disabilities want access to challenging landscapes, just like those who
may have no physical impairments. He invented what I refer to as a "nutritional food label" for trails, which provides essential information about
every trail, including surface (firm, soft, and material type), trail width,
inclination (running or longitudinal slope), cross slope, length, change
in elevation, and unique obstacles, such as the trail may become slippery
when it is raining. Providing this information at the point of entry to an
accessible pathway enables the user to make a decision on their desire
and ability to use a trail on any given day. It also provides trail designers
with latitude for layout and design of accessible paths.

The Grand Canyon Greenway is a challenging trail for everyone to use.
Visitors experience the effects of walking and bicycling at high altitudes,
with the South Rim at almost 7,000 feet and the North Rim at 8,000 feet.
The rocky terrain, uneven walking surfaces, and varying topography are
also challenges for accessibility. Led by Peter, our design team wanted
the greenway to link together popular destinations, some of which were
already connected by worn footpaths, but others that required new trail
routes and alignments. Building an accessible greenway required careful
route planning, trail design, and construction. Not only did we consider

every aspect of Peter's nutritional food label approach, but we also provided for strategically placed rest stops that enabled canyon rim trail users to stop and enjoy the view.

Over a period of ten years (1997–2007) the contract for design services grew to encompass six separate phases of greenway development on the South and North Rims of Grand Canyon National Park, totaling approximately seventeen miles of completed greenway trail. The following offers a brief overview of each phase of the greenway and some of the key aspects that factored into trail development.

Phase 1: Rim Trail: Yavapai to Pipe Creek Vista

The first phase of the greenway was in essence a pilot project, approximately 2.5 miles in length, and designed to link two of the most popular South Rim overlooks at Mather Point and Yavapai with a 10-foot-wide asphalt paved pedestrian pathway. The proof of concept had focused on this route and the construction details necessary to build the trail. Prior to greenway trail development, these two popular overlooks (Mather and Yavapai) were not formally connected and park visitors had established a series of braided trails across the rim top high desert landscape. The impact to the native landscape was severe, so our team and the National Park Service included landscape restoration as part of project development.

For most of the rim top greenway trails, we used asphalt as the trail tread surface. Originally, our recommendation was widely criticized within the Park Service and by Grand Canyon National Park advocates for not being environmentally sensitive. However, we had very good reasons to recommend asphalt as the tread surface. Foot traffic on the high desert landscape transformed the surface rock, known as the Kaibab Limestone layer, into a fine powder. The pounds per square inch of a heel print, when repeatedly applied in the millions across the same surface year after year, is enough to eventually crush the soft limestone rock layer. The limestone surface of the high desert landscape cannot support ten million heel prints without resulting in severe erosion. An asphalt trail provided us with a firm, stable, affordable, and workable wearing tread, critically important for accessibility during all types of weather. Asphalt was also a surface that could be installed and repaired by Grand Canyon work crews. The Park Service decided that it would use internal work crews to build the trail rather than contractors, due to the sensitive nature of the work. These crews were familiar with asphalt construction, and they did a remarkable job of building the trails.

The first phase of the greenway was constructed by these crews. In keeping with our design program, the asphalt trail was constructed so that it blended in with the surrounding high desert landscape. Park Service crews added rock walls to guide visitor travel. The Park Service also undertook an extensive revegetation program and eliminated the network of braided foot trails. In the end, everyone was pleased with how the trails looked and fit with the landscape.

Phase 2: Village Trail: Navajo Drive to Canyon View Information Plaza

The purpose of the second phase of greenway development was to build a two-mile trail from the Grand Canyon railway station on Navajo Drive to the proposed Canyon View Information Plaza, now called Grand Canyon Visitor Center. This phase was the equivalent of an urban trail, located through the heart of Grand Canyon Village. The 10-foot-wide asphalt trail offers park employees and visitors the opportunity to walk and bike to some of Grand Canyon Village's most popular destinations, including El Tovar, the railway station, Grand Canyon Park Headquarters, Park Services, Yavapai Lodge, the Grand Canyon Post Office, the Park Store at Grand Canyon Visitor Center and later the Bright Angel Bicycles rental shop and Mather Point Café.

Phase 3: Tusayan to Canyon View Information Plaza

The most ambitious and controversial phase of the greenway was the third phase, designed to link Grand Canyon Village to the town of Tusayan, Arizona, south of the park entrance. The project was more than seven miles in length and was controversial because the route of the trail would follow the original South Rim park entry road (known as the "two-track"). Today this trail is called the Tusayan Trail.

The challenge of phase 3 was designing a trail to follow an existing disturbed corridor. The project was subject to extensive cultural and natural heritage review. The National Park Service did not originally want to pave the trail, as was done with phases 1 and 2. However, some years later it was paved entirely within the park. Some sections had to be elevated on causeways (sections of trail elevated above the surrounding grade using rock walls to create an accessible travel route). Today, the trail links visitors from the heart of the South Rim to the Village of Tusayan where tourists can visit the IMAX Theater, stay at a number of hotels, and eat at a variety of restaurants.

Phase 4: Bridle Trail: Grand Canyon Lodge to North Kaibab Trailhead

The only phase of the greenway constructed on the North Rim was designed to link together the Grand Canyon Lodge, North Rim Campground, North Rim Visitor Center, and the North Kaibab Trailhead. The National Park Service refers to this as the "bridle trail," which was originally a dirt trail built to transport tourists on horseback from the Lodge to the North Kaibab trailhead. The unpaved improved greenway trail made the route fully accessible through a series of elevated causeways built across existing ravines and by widening the original bridle path to ten feet to support two-way travel.

Phase 5: Rim Trail: The Overlooks to the South Kaibab Trailhead

Phase 5 of the greenway is a connector trail built along the South Rim and Desert View Drive from the terminus of Phase 1 to Pipe Creek Vista, terminating at the South Kaibab Trailhead. Building this phase enabled park visitors to walk (or run) from Grand Canyon Village on the South Rim to the Grand Canyon Lodge on the North Rim, using the Kaibab Trail to complete rim to rim travel. The phase 5 trail is a ten-foot-wide asphalt paved trail that narrows in places due to constrained right of way.

Phase 6: Grand Canyon Village to Hermit's Rest

The final phase of the greenway links Grand Canyon Village to Hermit's Rest. This route of travel is one of the most rewarding sections of the greenway as it links together outstanding overlooks and transports visitors to more secluded and less traveled portions of the park. The trail extends more than six miles from Grand Canyon Village and includes the following destinations: Hermits Rest Viewpoint, Pima Point, Monument Creek Vista, The Abyss, Mohave Point, Hopi Point, Powell Point, Maricopa Point to Bright Angel Trailhead, and the Grand Canyon Village.

Millennium Trail Dedication

The Grand Canyon Greenway was formally dedicated in April 1999 at a ceremony that featured First Lady Hillary Rodham Clinton. The greenway was one of the most visible elements of the National Millennium Trails program, promoted by the Clinton White House as one way to celebrate the start of a new millennium and recognize America's rich trails history.

> These millennium trails will be very tangible gifts to the future. We will be able to walk on them and hike on them and bike on them. They will be accessible to people of all ages. But in a very important way they represent more than the tangible effects of the trail. They represent a commitment and an investment in the kind of country we want in the next century.[8]

Grand Canyon Greenway–A Model for Other National Parks

Today a visit to the South Rim or North Rim of Grand Canyon National Park would reveal nothing to suggest that six phases of greenway were stitched together over the course of ten years to form a continuous path of travel along the rims of the canyon. The asphalt trails blend with the other park infrastructure, providing seamless travel that is enjoyed by millions of visitors each year. Most important, the greenway has expanded accessibility across the South Rim to locations where users would not have dared to venture years ago and provided an important accessible path of travel on the North Rim that enhances the visitor experience. Our team set out to expand the time visitors spend at Grand Canyon National Park into a more immersive experience. Now it is easy to venture away from crowded overlooks, experience solitude, listen to the sounds of nature, and appreciate the awe-inspiring views and landscapes of the Grand

The Grand Canyon Greenway Collaborative is photographed with First Lady Hill-ary Clinton at a ribbon-cutting ceremony along the South Rim of Grand Canyon National Park. Source: The White House.

Canyon, and to accomplish this on foot, by bicycle, and by visitors with a range of abilities.

The Grand Canyon Greenway changed the way visitors experience a national park and is a model for other national parks to emulate. The rim-top greenway is not a traditional or typical solution; however, it is a replicable solution that can be applied in other national parks. The key takeaways include:

Connecting popular hubs of activity and destination landscapes within a national park with a network of improved and accessible trails. The ac-tivity areas and destinations can include campgrounds, concession areas, scenic overlooks, lodging facilities, restaurants, car parks, recreational vehicle campgrounds, and backcountry and wilderness trailheads.

Using a greenway solution to create a linear park along a resource edge. It was easy to accomplish at the Grand Canyon, but other national parks can examine their interpretive program and use a greenway to transport users along the edge of a resource for viewing and engagement. Trail builders can also use boardwalks and bridges to access sensitive landscapes without endangering the resources that people want to see and experience.

The Grand Canyon Greenway disperses the crowd and offers visitors a chance to engage with the natural and cultural resources of our national parks in a more intimate setting. This is a feature of the greenway that should be replicated in other national parks.

The greenway complements the national park transit system. The greenway connects with two major transit hubs at Grand Canyon National Park, offering visitors a choice in walking, bicycling, and the bus system to traverse the park.

Most important, the greenway provides accessible paths of travel for bicyclists and pedestrians. Not only does this address access for visitors with varying physical abilities, it also organizes visitor foot and wheeled travel within the park, limiting the number of braided trails and their impacts to adjacent landscapes.

The greenway has provided visitors with a more meaningful connection to the unique ecology of the Grand Canyon high desert landscape. The landscape is fragile, and five million people visiting the South Rim is a significant impact to ecology, biodiversity, and plant and animal habitat. The greenway offers access to this landscape so that it can be appreciated for its intrinsic value, stewarded, and conserved for future generations to enjoy.

Greenways are not the traditional backcountry wilderness trails or simple footpaths within our national parks. Greenways have different purposes and design criteria; they serve a unique function that enhances the visitor experience. Greenways can be the perfect solution to creating paths of travel for the large crowds that make use of our National Parks.

Bicycling in the Park

One of the most significant achievements of the project is the ability to use a bicycle for travel throughout the park. Bright Angel Bicycles, a vendor located at the Grand Canyon Visitors Center, rents bicycles for hourly or daily use. With more than twelve miles of rim-top trails along

A cyclist enjoys a section of the Grand Canyon Greenway along the South Rim of Grand Canyon National Park. Source: Sarah Neal, Bright Angel Bicycles.

the South Rim, the use of a bicycle provides an efficient, human-powered transportation choice and encourages visitors to move away from crowded overlooks and explore the canyon at a self-determined pace. If you choose to rent a bike or bring your own, exercise caution as you ride off on your adventure—oxygen levels are lower at the South Rim, and it can be easy to overexert in the rush of excitement and joy of the bicycle ride.

On a personal note, I realized the very minute that I set foot into Grand Canyon National Park that whatever time I spent working in the landscape would be one of the most unique and exceptional professional experiences of my career. I am in total agreement with President Roosevelt that there was nothing we would ever accomplish that could improve upon the natural beauty of the park. That was never our intention. All we wanted to do was provide a better mode of safe, accessible travel that everyone could use to enjoy and experience the fullest extent of the Grand Canyon. The Grand Canyon Greenway provides vital park infrastructure that engages the body and mind of each visitor and reinforces the sense of place that makes the Grand Canyon one of the most unique and most visited landscapes on the planet. Hopefully, Roosevelt would agree that we did not "mar" the landscape but, given the conditions of intense visitation, made appropriate changes that enable more parents and their children to have improved access and therefore greater enjoyment of this magnificent wonder of the world.

5

Open Space in Vegas—It's a Sure Bet

Las Vegas Open Space and Trails, Las Vegas, Nevada

Tule Springs Lake in Floyd Lamb Park, Las Vegas, Nevada. Photo by author.

IT IS KNOWN BY SEVERAL NICKNAMES: "Sin City," "Glitter Gulch," "City of Lights," "Gambler's Paradise," and the "Entertainment Capital of the World." Las Vegas is a city unlike any other on planet Earth: an adult playground; a city that never sleeps; and a place where tourists go to unwind, release inhibitions, and celebrate their fantasies. On closer inspection, when you examine Las Vegas beyond the casinos, entertainment shows, the all-you-can-eat buffets, neon lights, and tall buildings that line Las Vegas Boulevard, you reveal a city of people who live in a variety of neighborhoods, purchase their groceries at local supermarkets, attend local schools and colleges, and work in various office buildings scattered across the sprawling metropolis. Las Vegas in this sense is very much like other American cities. Replace gambling with the oil and gas industry, and you have Houston, Texas. Substitute gambling for higher education, technology, and the pharmaceutical industry, and you find yourself in the Raleigh-Durham metropolitan region of North Carolina. Gambling and entertainment are the economic engine of Las Vegas, and yet the needs of her citizens are very much the same as they would be in similar-sized American cities and metropolitan regions.

Las Vegas Means "The Meadows"

Las Vegas translated from Spanish means "fertile plains." Another more widely accepted translation is "the Meadows," which is derived from the fact that in certain portions of the Mojave Desert spring water surfaces and along with a high water table supports a diversity of grasses, cactus, and flowering plants.

In the late 1800s, Las Vegas was known to have provided crucial support for westbound Union Pacific steam engine trains heading to California that needed to fill their water tanks so they could climb through the Sierra Nevada mountains. The most suitable place in the Mojave Desert to accomplish this important task was at one of two naturally occurring springs just east of the mountain range. One of those springs was named Las Vegas and the other Tule Springs.

Settled in 1905 and incorporated in 1911, Las Vegas is one of the fastest-growing cities in the world, now the twenty-eighth most populous in America, and one of the most visited tourist destinations. According to the City of Las Vegas Visitors and Convention Bureau, each hotel room constructed creates the need for 1.1 casino-related jobs and 1.9 service or support-related jobs. In 2018, Las Vegas offered more than 160,000 hotel rooms, the most of any city in the United States, and approximately 32 percent of the labor force of the city is employed by casinos.

When most people think of Las Vegas, what immediately comes to mind is the infamous Las Vegas Strip, miles of casinos and hotels along both sides of Las Vegas Boulevard. In actuality, the Strip is not located within the City of Las Vegas, but rather south of the city limits in unincorporated Clark County, Nevada. The City of Las Vegas proper is much older than the Strip. The legacy casinos that line Fremont Street, such

Aerial view of sprawling residential subdivisions in the Las Vegas Valley. Source: Author.

as Binion's Gambling Hall, the Golden Nugget, Fremont, and the Four Queens, are no longer the most popular destinations in the metro area.

As Las Vegas grew, much of the land development occurred with little regard for the park, greenspace, and open space needs of its citizens. That began to change in November 2002, when the City of Las Vegas hired my company to prepare an Open Space Plan for the city's northwest region. The focus of our work was not the Las Vegas strip, nor was it the historic downtown area. Northwest Las Vegas is the sprawling suburbs, miles north of the Strip and downtown, where people who worked in the casinos and the service and support sector live.

Soon after my contract had been signed, I was questioned by Las Vegas Deputy City Manager Steve Houchens about the need for an open space plan. Steve reminded me that "the City of Las Vegas is surrounded by millions of acres of open space, in every direction, for as far as one can see. All I need to do is get in my car, drive 30 or 40 minutes from downtown in any direction, to the edge of the city, and I am surrounded by vast acres of open space." Steve asked me point blank: "Why have WE hired YOU to prepare an OPEN SPACE plan for the City of Las Vegas?"

I fully understood both the context and motivation for Steve's question. Why do we need undeveloped open space, parkland, and greenways within American cities, especially when there are vast rural landscapes that surround many American cities? What are the benefits that open space provides? Why should cities like Las Vegas invest time and money in preparing open space plans? I realized that if I could successfully answer Steve's question, I would in fact be able to address a broader audience of 1.2 million potentially skeptical and curious residents within one of the world's fastest-growing cities. It was a critically important question to ask and answer, and my team of landscape architects and urban planners needed to sharpen our focus and deliver a compelling argument for conserving and protecting the native Mojave Desert landscape and rapidly dwindling open spaces of the Las Vegas Valley. Steve's question was a defining moment during our work and eventually served as a foundation for the first open space master plan in the ninety-one-year history of Las Vegas.

The Legacy of Olmsted and Howard

The idea that cities should set aside land, plan for, and include parkland and healthy outdoor living spaces for residents can be attributed in part to Frederick Law Olmsted and the creation of Central Park in New York City. America's park and recreation movement got its start in the 1850s when city leaders living in densely populated cities began to understand the link between overpopulation, failing public health, and lack of access to healthy outdoor space. New York City was the most populated American city in the mid-1800s and decided to sponsor a competition for the design of a 900-acre tract of land that would clean out a squalid, overpopulated, unhealthy landscape in the middle of Manhattan island and replace it with a park. The competition resulted in the 1858 "Greensward" plan—now known as Central Park—submitted by Olmsted and his partner Calvert Vaux. Neither were park and greenspace planners per se. Vaux was a noted British-born architect, and Olmsted had no formal professional title at the time of their submittal.

Greensward means by definition "an area of ground that is covered by green grass." Olmsted and Vaux's Greensward was much more than a master plan for green grass and trees in New York City. The designers made a deliberate effort to organize and program outdoor space in a way that would support specific activities and encourage interactions among park users. One of the great legacies of Central Park is that it is "democratic space" regardless of one's economic status, often referred to as the "people's park." Every element of Greensward had an intended outcome and was thoughtfully designed to produce healthier and happier people.

With the success of Greensward and the construction of Central Park came the desire for every major city in America to begin planning, designing, and constructing their own version of Central Park. This eventually led to the establishment of landscape architecture as a profession and the formation of the National Park System as an expression of America's desire and intent to conserve and protect scenic and valued native lands and waters across the nation.

Forty years after Olmsted and Vaux submitted the Greensward Plan, the Garden City Movement began in England under the leadership of Sir Ebenezer Howard, one of the first urban planners to put forth the idea that healthy workers needed to live in healthy cities, and healthy cities in turn required significant amounts of greenspace, consisting of parks, conserved woodland, gardens, and other natural undeveloped open spaces. Howard's visionary work was labeled utopian at the time because there were no actual cities constructed as he envisioned. He put forth a concept plan for a "Garden City" consisting of concentric rings of open space and parks with connecting radial boulevards that could house 30,000 people on approximately 6,000 acres of land. Howard's ideas generated discussion among city designers, architects, politicians, and city leaders about the proper mix between developed and undeveloped land within cities. Las Vegas was founded a decade after Howard expressed his utopian vision for city development, and as the city grew it contained none of his ideals for the inclusion of open space, parks, and shaded boulevards.

Saving the Leftover Land

The efforts of Olmsted and Howard helped define a new approach to city planning, design, and construction. However, during much of the twentieth century most American communities developed and grew without regard for greenspace conservation or for adequate provision of park and recreation lands and facilities. Open space conservation typically has been an afterthought and a poorly executed element of the land use planning process. Historically, community planning, as conceived and codified by the American Planning Association, and implemented in similar fashion by the American Society of Landscape Architects, has focused on accommodating population growth and land development and transforming undeveloped rural land and greenspace into housing, shopping centers, schools, industrial complexes, office parks, highways, and institutional uses. This was clearly the framework for the growth of Las Vegas. The goal was to build on the expanse of available desert land as quickly as possible and to make as much profit as could be realized in the process. Open space, in this growth and development process, was the leftover land, often the land that developers could not figure out how to build on, or the land on the edge of the city yet to be built on.

From my perspective, open space conservation and protection is one of the hallmark achievements of landscape architecture. Among Olmsted's signature accomplishments were his contributions to open space preservation, stewardship, and resource management. Early in my career, I recognized that open space was more than scraps of discarded and worthless parcels of land not suitable for human or economic purpose. I advocated for open space as green infrastructure and felt that we should be planning and designing open space systems, much like we planned and designed our gray infrastructure systems (water, sewer, road, and other utility systems). I developed a summary statement to frame the argument: "Just as we would not consider building a home without a set of blueprints, we should not consider growing and building our communities without a greenprint—a green infrastructure plan that determines the valued open space we should conserve and protect as part of the growth and development process."

Putting the need for open space conservation into terms that community residents could understand worked. It resonated with people from all walks of life. Open space should not be the by-product of the land development process; it should be integrated into the process. We should engage in a community conversation with residents to identify, discuss, and plan for the lands that they value and therefore are deserving of stewardship, conservation, and protection. That is exactly what we accomplished in the City of Las Vegas with the open space plan, and it changed the perception regarding the need and desire to protect the native desert ecosystems within the city limits.

Green Infrastructure in the Mojave Desert

According to Mark Benedict and Ed McMahon, authors of *Green Infrastructure*, these protected and conserved resources provide "an interconnected network of green space that conserves natural ecosystem values and functions and provides associated benefits to human populations. The ecological framework needed for environmental, social and economic sustainability—in short it is our nation's natural life sustaining system."[1]

Green infrastructure consists of native vegetation, such as forests, meadows, and grasslands; stream corridors, wetlands, riparian lands, and floodplains that transport and purify surface waters; soil; and a variety of landforms. Green infrastructure offers many benefits to communities

A residential subdivision trail in Las Vegas, Nevada. Photo by author.

including management of drinking water supply and absorption of rain water; purification of our air and mitigation of the harmful pollution; conservation of landscapes that moderate the temperatures of our communities, offsetting the urban heat island effect that results from densely developed landscapes; sequestration of carbon in our soils so it can give life to trees and vegetation; enhancement of economic value for adjacent public and private property; and protection of homes and businesses from extreme heat and cold, thereby lowering energy consumption.

Conversely, gray infrastructure is the human-designed, engineered, and constructed systems inclusive of roads and highways, utility corridors, water and sewer systems, telecommunications systems, and the like. Without question, the development of gray infrastructure has advanced human civilization, enhanced the quality of life in our communities, improved the standard of living, and extended human lifespan. Both gray and green infrastructure are vitally important to the quality of life in communities.

In the Mojave Desert, where there is very little "green," how would the concept of green infrastructure apply to the City of Las Vegas? The desert is a more delicate and fragile landscape generally lacking sufficient water, weathered over thousands of years, but still capable of supporting a diverse array of plant and animal life. While the native open space infrastructure is not green, as it would be with more rainfall and forest canopy cover, it is nevertheless a substantial ecosystem that is worthy of conservation and protection.

The Start of a Green Infrastructure Initiative

The fifty-square-mile northwest region of Las Vegas has been one of the fastest-growing areas in the United States since the 1980s. It was important to understand the devastating impact that rapid growth was having on the native desert ecosystem and rural character of the region.

For the Northwest Las Vegas Open Space Plan, my team included local landscape architect Jack Zunino, FASLA, my friend and colleague Bob Searns from Denver, Colorado, and the firm of ETC Leisure Vision based in Kansas City, Missouri. Jack Zunino was critical to the success of the project. Born in Nevada, Jack is to this day one of the most significant landscape architects in the history of Nevada. Jack earned a bachelor's degree in psychology at the University of Utah but then switched gears and obtained a master's of landscape architecture and environmental planning at Utah State University. Having dual degrees in psychology and landscape architecture has most likely enabled Jack to enjoy staying power and success in the Las Vegas market. Jack and I were inducted in the same American Society of Landscape Architects 2003 Class of Fellows, an award that recognizes a lifetime of professional achievement. We never met during the induction ceremony in New Orleans, but when I was searching for a landscape architect to partner with in Las Vegas, I remembered his name, realizing we were in the same fellows class. I figured that some sort of cosmic energy was bringing us together. I was correct. Jack operates a small and yet very successful practice in Las Vegas, opened for business in 1989, which continues operation to this day.

Bob Searns is an architect and planner whom I have already described in chapter 4 of this book. It is worth repeating that he is one of my best friends and colleagues in the business. Working together in Vegas was more fun than either of us could have imagined. Bob provided a western United States perspective to our planning and design work that was critical to the success of the project. He also worked extremely well with Jack Zunino, as well as the entire client team, and would at times be the lead representative for our team at client meetings. Among Bob's favorite recollections were the many times that I would travel to Las Vegas to conduct work. I typically traveled to Las Vegas once a month, and there is a three-hour time difference between my home in Durham, North Carolina, and Las Vegas. Bob likes to play poker and would usually arrive in Las Vegas the evening before I was scheduled to arrive in order to enjoy a few hands before bedtime. Normally around 1:00 a.m. Vegas time, Bob would fold up his last hand and head to his hotel room. Checking his watch, he would say to himself, "Chuck is just getting out of bed and heading to the airport to catch his flight to Vegas, which gives me just enough time for a good night rest." All of which was true. I normally caught the 5 a.m. flight out of Raleigh-Durham and would arrive in Las Vegas before noon, ready to commence work for the rest of the day and evening. Working with Jack and Bob was rewarding personally and professionally, and the three of us combined our talents and expertise to generate innovative solutions to the challenges of open space planning in Las Vegas.

The City of Las Vegas team was led by Tom Perrigo, who from 2003 to 2005 was planning manager and today serves as executive director of community development and chief sustainability officer for the city. Tom was a champion for conservation and green infrastructure long before it became popular nationally. He is a critically important leader, not only in the City of Las Vegas, but also throughout Southern Nevada, with respect to open space protection, conservation, and stewardship. It was an honor to work alongside Tom for six years as together we addressed critically important land conservation projects. Tom's influence was included within every plan, program, and policy statement we produced.

To generate our recommendations, my team produced time-lapsed geographic information system (GIS) mapping to illustrate the rapidly changing land use during the prior ten-year period. We analyzed natural heritage areas, remaining significant open space, arroyos and washes, as well as settlement patterns. The most interesting aspect of this analysis was the lack of significant large-scale natural features that could serve as the basis for establishing a framework for the open space plan. Aside from the Las Vegas wash and surrounding mountain backdrop, the northwest region of the city was an endless expanse of sand and rock. The desert floor of the Las Vegas Valley seemingly had little redeeming natural value, and because of this, both government and the private sector had little regard for conserving this native landscape. It was just another expanse of sand that would be transformed into residential, commercial, or industrial uses.

Defining the Need for Open Space

A turning point during our planning process occurred when we completed and published the results of a Community Attitude and Interest survey that was conducted by project partner Ron Vine, vice president of ETC/Leisure Vision.[2] It was a very comprehensive survey, and it provided the basis and philosophy for the Open Space Plan. Significantly, the survey confirmed for our planning team and the city the objectives of the plan. The residents of Northwest Las Vegas expressed their support for the conservation of the native Mojave Desert ecosystem and wanted to protect natural areas and viewsheds, habitat of plants and animals, the washes and other areas that were prone to flash flooding, while at the same time providing more areas for trails and landscapes that supported health and fitness. Las Vegas residents' desire for new sports fields, golf courses, and buildings, such as gymnasiums and community centers, was a low priority in the citizen survey. The highest-ranked activity for Northwest Las Vegas residents was walking and hiking, followed by wild-life habitat viewing.

The results of the survey stunned some members of the city staff. Even though my agreement with the city specified preparation of an open space plan, there were some city staff who firmly believed that active parks and recreation would be foremost on the minds of their citizens. The survey results also surprised Mayor Oscar Goodman and the Las Vegas City Council. The ETC/Leisure Vision survey results were critical in guiding the preparation of the Open Space Plan.

Oscar Goodman was the mayor during the six years that I worked in Las Vegas. Born and raised in Philadelphia, Mayor Goodman is best known as being the lawyer for the Las Vegas mob, who then went on to become mayor and "First Gentleman" of the city. As an accomplished defense attorney, Goodman represented some of the better-known members of organized crime in Las Vegas, including Nicky Scarfo, Herbert "Fat Herbie" Blitzstein, Frank "Lefty" Rosenthal, and Jamiel "Jimmy" Charga. One of his most famous clients was Anthony "Tony the Ant" Spilotro, who was depicted as Nicky Santoro (played by actor Joe Pesci) in the 1995 movie *Casino*. Goodman was used to leadership roles, once serving as president of the National Association of Criminal Defense Lawyers in 1980–81. Most important, Oscar Goodman loves and has thrived in Las Vegas. He is still regarded as the city's most popular mayor, having served three four-year terms from 1999 to 2011. His wife Carolyn succeeded him as mayor in 2011 and still holds that title today. Mayor Goodman's memoir is called *Being Oscar: From Mob Lawyer to Mayor of Las Vegas*.

One of the most memorable exchanges with Mayor Goodman happened when I stood at the podium to introduce the final Northwest Las Vegas Open Space Plan. Goodman feigned ignorance that Vegas had a northwest region of the city (despite the fact that it is the largest geographic region of the city). Playing to the council chamber audience, Goodman challenged the citizens to help him find the northwest region.

One clever citizen provided the perfect response: "Mayor, use your nose and follow the scent north up Rancho Drive, when you step in 'it,' you will know that you have arrived," a reference to the fact that Northwest Las Vegas was home to hundreds of equestrians, many of whom boarded their horses in backyard makeshift sheds, stables, and corrals. Mayor Goodman and the entire Las Vegas City Council were impressed by our work and unanimously approved the Northwest Open Space Plan. The planning department staff later shared that it was the first time they could remember a unanimous decision by the mayor and city council.

Establishing a Foundation of Support for Conservation

The final Open Space Plan targeted programmed and nonprogrammed open space devoted to the conservation of natural resources, outdoor recreation, preservation of historic and cultural property, protection of scenic landscapes, and protection of public health, safety, and welfare. The plan combined our technical evaluation with the results from the public survey and other input to define specific recommendations.

In the absence of clear and definable natural features to anchor the open space plan, such as rivers, lakes, forests, or other easily discernable natural features, the plan defined a set of broad goals and objectives to guide city decision-making. First, a goal of protecting 30 percent of the remaining undeveloped land in the northwest region as future open space was defined. This bold declaration provided a target for open space conservation that was based on the ecological, social, economic, and political realities of Las Vegas and was designed to ensure a quality of life for city residents that is progressive, sustainable, healthy, and economically viable. The plan recommended the conservation of approximately 4,000 acres of land in the Northwest Las Vegas region.

Second, the open space system was based on a "hub-and-spoke" concept that would help link community residents to employment centers, shopping areas, parks, and other popular destinations. Hubs would be parks and open space conservation areas; spokes would consist of off-road greenways and on-road trails for walking and bicycling. Four components illustrated the open space network map, including (1) protected native lands, (2) active recreational landscapes, (3) historic and cultural landscapes, and (4) contiguous open space corridors.

Third, in addition to the hub-and-spoke network, the plan recommended the stewardship of federal and state lands that surround the Las Vegas Valley, which we referred to as the "Las Vegas Vias Verdes." The Vias Verdes was envisioned as a large-scale conservation strategy for publicly and privately owned lands that surround the valley at the base of the mountain backdrop. The lands within the proposed Vias Verdes would include Lake Mead National Recreation Area, portions of the Sheep Mountain range, Sloan Canyon National Recreation Area, Desert National Wildlife Refuge, Red Rock Canyon National Conservation Area, and Lake Mead. This was a new concept for federal land managers and local governments in Southern Nevada and was enthusiastically supported by the city staff and residents.

The most important objective of the plan was getting the city to identify key parcels of land for conservation, which was difficult to accomplish at that time due to the fierce competition for land. We made every effort to emphasize that conservation can only happen when the City of Las Vegas takes an aggressive approach, which was not necessarily the highest priority of city government. We felt that this aggressive approach was needed as land costs were too expensive to realize successful open space conservation through "voluntary" participation. Therefore, the plan contained a variety of strategies, including (1) revisions to the land development regulatory framework to encourage more open space conservation; (2) a variety of land acquisition tools to protect land before it is developed; and (3) the use of federal, state, and local funds to purchase land.

One Plan Becomes Four

The City of Las Vegas was pleased with the way that my team had crafted a master plan for open space conservation and immediately saw opportunities to build on the success of the plan. In June 2004, the city amended the original contract to include a trails master plan as a logical next step in further defining the hub-and-spoke framework of the Open Space Plan. During the next four years, the city amended the contract three more times. Altogether, I spent six consecutive years working on a variety of open space and green infrastructure projects that included the Southern Nevada Regional Open Space Plan and a master plan for Floyd

Lamb Park at Tule Springs. The culmination and combination of this work provided a solid answer to Steve Houchens's question, "Why do we need an open space plan?"

City Trails Plan

In the spring of 2004, we began work on a 200-mile network of shared-use, transportation-oriented trails, complete with trailheads, that would support nonmotorized (bicycle and pedestrian) travel across the city. The primary emphasis of this City-wide Trails Master Plan was ten trail corridors, all of which were named for the roadways that they followed. Due to historic land development patterns of the city (again Las Vegas did not develop in a manner that Ebenezer Howard would have preferred), roadway corridors and electric transmission lines were the only available linear landscapes extending across the city available for trail development.

The City of Las Vegas had been awarded grant funding from the Southern Nevada Public Land Management Act, known by its acronym "SNPLMA." This program was begun in 1998 and provided a process whereby the United States Bureau of Land Management could sell federal lands located within the Las Vegas valley for private development (subdivisions, retail, commercial, industrial, and institutional development). The sale of federal lands was one of the primary reasons that Las Vegas was growing so quickly. Through SNPLMA there was a seemingly endless supply of land to satisfy the growth (almost 7 million acres of open space). The revenues derived from the land sales were divided between the State of Nevada General Education Fund, the Southern Nevada Water Authority and a special account managed by the Secretary of the Interior that could be used to support a wide variety of green infrastructure projects, including parks, trails, and conservation landscapes. It was the third category that provided the city with funding for trail development.

We completed the trails work program within six months (August 2004 to February 2005) using geographic information system technology to identify and name the trail corridor segments, evaluate gaps in the overall trail network, plan and design trail connections within those gaps, prioritize a roster of trail construction projects, and provide the city with construction specifications and estimates for building the network.

Two women hike along an existing trail in the Las Vegas Valley. Photo by author.

The city awarded construction contracts immediately and began transforming the targeted corridors into finished trails, which were combined to form an interconnected network. This was the most successful element of the open space and trails planning and design work in Las Vegas in that all the trails were constructed as planned and designed.

Regional Open Space

In November 2004, the Southern Nevada Regional Planning Coalition (SNRPC) hired my team to complete an open space plan for all of Southern Nevada. The focus of this work took into consideration the needs of large cities, like Las Vegas and the city of Henderson, as well as smaller communities including Boulder City and the city of North Las Vegas.

We worked with a coalition of municipal park and recreation directors, which later became known as the Southern Nevada Regional Coalition Open Space and Trails Workgroup (ROST), to better understand the challenges of rapid population growth and its impacts on the landscapes of the region. There was an urgent need for a strategy to preserve and protect open space before it was lost to rapidly occurring land development. What SNRPC needed was a vision for regional open space conservation and trail development, along with tools and a framework for open space conservation, which we developed and adapted from the Nevada Revised Statutes (NRS 376A.010). We concluded that "open space" should include the Mojave Desert, mountains, special geologic and topographic features, meadows, wetlands, washes, lakes, working agricultural and ranch lands, and other valued landscapes and ecosystems.

One of the first efforts of my team was to evaluate prior work and identify regionally significant open space resources that we felt were either threatened or worthy of conservation. This effort produced a targeted 96,000 acres of open space located throughout the Las Vegas Valley.

The Southern Nevada Plan also identified key landscapes that should be part of a regional open space system, to encourage the conservation and protection of tens of thousands of acres of valued open space. These key landscapes included:

The mountain and desert backdrop that was critical to the identity of the Las Vegas Valley.

The Vias Verdes, originally identified in the Northwest Las Vegas Open Space Plan, a transitional belt of lands that are located at the base of the mountain range, encircling the valley and regionally significant for their ability to provide a sizeable buffer between the urbanizing valley floor and the protected mountain range.

Washes and arroyos; landscapes that flood during rainstorms and that should not be developed with housing or commercial, retail, or institutional land uses.

A regional trails network that would be developed across the valley floor, connecting communities to significant regional destinations.

Regionally significant heritage open space, which included archaeological and paleontological landscapes valued for their scenic, natural, and cultural qualities.

These five regional landscape elements were the skeletal framework for the Southern Nevada open space plan and were illustrated on a map to depict the regional open space system. Each of the landscapes contained specific planning and management objectives, and a conservation toolbox was proposed to provide state, regional, and local governing strategies for conserving and protecting these resource areas.

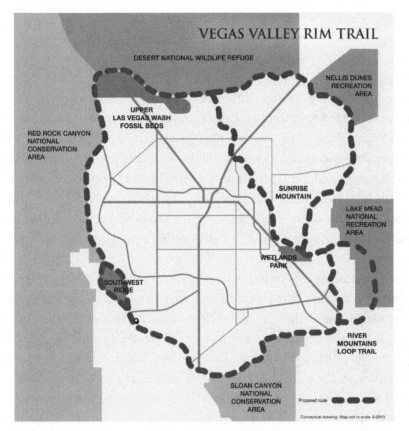

Map of the Vegas Valley Rim Trail. Source: Get Outdoors Nevada, Southern Nevada Regional Planning Coalition, and Regional Open Space & Trails Workgroup.

We concluded our work outlining a series of "next steps" to guide implementation of the plan recommendations. The conclusions of our work were endorsed and acted on by a regional conservation organization, the Outside Las Vegas Foundation, that today is known as Get Outdoors Nevada. This organization began in early 2000 as an effort to connect the people of Southern Nevada with outdoor places that residents and visitors alike enjoy. One of the most tangible recommendations from the Southern Nevada plan was a 113-mile Vias Verdes (originally conceived as part of the Northwest Las Vegas Open Space Plan), which Outside Las Vegas transformed and renamed the Vegas Valley Rim Trail.

Saving Tule Springs

The final open space planning contract that I undertook for Las Vegas began in the spring of 2006 when the City asked if we could help them save historic Tule Springs from development. Targeting a specific tract of land enabled my team to transform vision into action and resulted in Las Vegas taking ownership and management of Floyd Lamb State Park, where Tule Springs is located. The State of Nevada was proposing to sell the historic parkland to private developers in order to generate funds that would help erase a state budget deficit. The city strongly objected to the state's decision and entered into an agreement to take control of the state park and surrounding lands. The city needed a master plan for the park to seal the deal.

Complicating the transfer of land to the City of Las Vegas was the fact that there was no city department or agency that would assume management of the park. The City of Las Vegas did not have a traditional park and recreation department and instead operated a Leisure Services department, which was primarily involved with programming events and activities on city-owned parkland. Management of parkland was distributed across other agencies in the city. We were charged with authoring a governance, operations, and management plan for Floyd Lamb Park that would define staffing, equipment and facilities, duties and responsibilities, and an annual operating budget.

Floyd Lamb State Park had a checkered history. It originally began as a community park in 1964 when the City of Las Vegas purchased 680 acres of land and water rights to Tule Springs Ranch. This lush oasis in the Mojave Desert was famously known as the "divorce ranch," established in the 1940s by entrepreneur Jacob Goumond. Married couples wanting to divorce could do so in Nevada after fulfilling a six-week stay at the ranch. Goumond built a collection of small cottages that could house between ten and twelve adults while they waited out the six-week timeframe required by Nevada law. Purchasing the land in 1964, the city sought to preserve the unique cultural heritage of the ranch and the natural features of Tule Springs, which was one of two prominent surface water features in the Las Vegas Valley, the other being the historic Las Vegas Springs, located fifteen miles south.

In 1977, Floyd Lamb, the Nevada State Senator who represented Clark County, convinced the State of Nevada to rescue the ranch and city park from financially troubled Las Vegas, and the city park was subsequently renamed Floyd Lamb State Park. In 1983, Lamb was convicted of accepting bribes in an FBI sting operation and was sentenced to a nine-month prison term. Ever since his conviction, Las Vegas residents have believed that the name "Floyd Lamb" should be dropped and favor referring to the park by its natural feature, Tule Springs. Today, the city park is named Floyd Lamb Park at Tule Springs.

Tule Springs is more famously regarded for the spectacular fossil beds of ancient animals located in a "big dig" area just north of the park. Identified in 1962 during an intensive paleontological excavation, the big dig was documented by the National Geographic Society and is thought to be one of the most significant paleontological sites in North America. The word "Tule" can be traced to Aztec Indians and means "group of plants" that would grow near water. Scientists believe that at one time, the Las Vegas Valley was a huge lake teeming with all forms of life among lush swaths of vegetation. Tule Springs and the Las Vegas Springs are the remaining elements of this ancient lake. The reason so many large animal bones have been discovered in the area is due to the presence of a large ancient lake bed. Fossilized remains, some more than 250,000 years old, of bison, camels, horses, mammoths, and giant sloths, along with the giant North American lion, have been found within the big dig area. In 1979, approximately 1,000 acres was added to the National Register of Historic Places, listed as the Tule Springs Archaeological Site. During the past decade, a coalition of Las Vegas residents, known as "The Protectors of Tule Springs," continued their quest to have the area protected and succeeded in 2014 with the federal designation of the Fossil Beds National Monument Act.

The study area for our master plan totaled more than 2,200 acres of land in Northwest Las Vegas and encompassed the Tule Springs Ranch, an existing skeet shooting range, a proposed NW Las Vegas Equestrian Park, and thousands of acres of Recreation and Public Purpose lands that had been leased from the Bureau of Land Management.

My team mapped and inventoried the Tule Springs Ranch. We provided a detailed historical record of the natural and human-made features

PREFERRED ALTERNATIVE

Preferred Master Plan for Floyd Lamb Park at Tule Springs. Source: Author.

of the site, including the historic ranch building complex, the spring-fed lakes, and other features that had been used by the public for decades. During our work, we discovered that no detailed inventory of the ranch existed, so we completed a map and written summary of the existing structures and their condition so that there would be a baseline of information for future use by the city and historians.

Las Vegas residents were vested in the process and the outcome of our work and attended our working meetings in large numbers to contribute to the design of the park. Three development alternatives were offered for the park that defined varying intensity of park facility development. A preferred alternative for park development featured a new park entrance road, visitors center, and archaeological museum (with artifacts from the big dig), a recreational off-road bike park (mountain bike park), a nursery for use by the Nevada Division of Forestry, upgrades to the park core area, and two environmental enhancement areas. The plan included recommendations from the NW Equestrian Park Master Plan, a network of trails throughout the Recreation and Public Purpose lands, and reserved a block of land for a future community park that would feature ballfields, playgrounds, and other amenities.

The north and south environmental enhancement areas were the result of a request from the Clark County Regional Flood Control District. Land development in Northwest Las Vegas has historically paid little attention to the fact that much of the Mojave Desert is subject to flash flooding during heavy downpours. The district had been working to mitigate the impacts of urban flooding and discovered that some of the only remaining undeveloped land in Northwest Las Vegas was in and around Floyd Lamb Park. They approached the city with a proposal to build two stormwater detention basins within the park. Since the transfer of land had not yet occurred, the city directed the district to work with the Nevada Division of State Parks.

No one was thrilled with the fact that two large stormwater detention basins were going to be built within the park. Steve Weaver, deputy director of state parks, asked me to assist him and provide guidance and leadership on the stormwater basin design. I believed that we could design the basins in such a manner that they would blend in with the surrounding desert landscape. The resulting construction achieved detention basins that, despite their massive size, were built in a way that created an amenity for residents while simultaneously satisfying the need for regional stormwater detention.

Las Vegas city councilman Steve Ross was a big champion of the completed work and was thrilled to cut the ribbon on the newly created detention basins. Today Floyd Lamb Park at Tule Springs is one of the top local attractions for visitors and residents alike. Ironically, it is one of the most requested venues for marriage proposals and wedding ceremonies.

Establishing an Ethic for Conservation

The combined efforts of the four projects—NW Las Vegas Open Space Plan, Las Vegas Trails Plan, Southern Nevada Regional Open Space Plan, and Floyd Lamb Park at Tule Springs master plan—created a foundation for launching a green infrastructure initiative throughout Las Vegas and across southern Nevada. Since the completion of work (in the year 2007), the level of interest in conservation and park and trail development has grown exponentially across the city and throughout southern Nevada. Outside Las Vegas Foundation, under the leadership of Alan O'Neill, took on a prominent role in leading the charge. The regional trails and open space working group (ROST) established for the southern Nevada plan took on the role of implementing recommendations and linking the communities of southern Nevada together with trails, parks, and conserved open space. There are other numerous achievements and accomplishments, including:

In 2013, the City of Las Vegas adopted a new comprehensive plan, "Vegas Master Plan 2020," which includes chapters on conservation, trails, and recreation that build on the work of the Northwest Open Space Plan and City Trails Plan, advocating for an interconnected network of open space, trails, and green infrastructure.

The Las Vegas Trails system has grown significantly. Originally envisioned as a network of 200 miles in total length, today more than 140 miles are on the ground and available for public use. In 2002, the city had 8.5 miles of constructed nonequestrian trails, and by 2012, more than 87 miles of these trail types were available for public use.

Likewise, the city's park system has also expanded tremendously. A network of modern parks is appropriately and strategically located across the city, offering residents and visitors access to a variety of outdoor venues, historic and cultural landscapes, and recreational activities. These green infrastructure hubs are linked by the expanded network of spokes, in the form of regional trails. The largest green infrastructure hub in the system is Floyd Lamb Park at Tule Springs.

The City of Las Vegas is taking steps toward becoming a bicycle-friendly community. In 2002, the City of Las Vegas had no on-street bicycle facilities, and by 2012, the city had developed 119 miles of bike lanes.

The regional network of trails in southern Nevada has grown substantially. Now envisioned as a 1,000-mile interconnected system linking together resources in Clark County, Henderson, Las Vegas, and North Las Vegas, it will feature a diverse array of hiking and bicycling trails.

Clark County's Health Department has worked in partnership with ROST and the Outside Las Vegas Foundation, supported by a Robert Wood Johnson Foundation grant, to launch "Neon to Nature," a regional trails program whose purpose is to connect Southern Nevada residents and tourists with the outdoors. Neon to Nature, which connects the gaming capital of the world to the spectacular natural assets of Southern Nevada, offers programs to promote walking, bicycling, and health and wellness initiatives. The program includes a mobile app that provides users with up-to-date information about the regional network of trails. ROST supported the development of a regional wayfinding and signage system so that users can better understand how to use the interconnected network of regional trails.

Managing stormwater from rainstorms continues to be a challenge for Clark County and the municipalities of southern Nevada; however, the template for multiuse facilities, as constructed in Floyd Lamb Park, offers alternatives to traditional engineered solutions.

The agenda for land conservation continues to grow in importance across southern Nevada. The pressure to grow and develop land also continues. In 2018, Clark County proposed the conservation and protection of 400,000 acres of land to offset proposals for developing an additional 38,000 acres of land throughout the valley (in essence 10.5 acres of conserved land for every acre developed). Clark County maintains that this initiative represents a balanced approach between conservation of open space and the need for development. Despite these efforts, continued sprawl into the undeveloped Mojave Desert remains one of the most controversial facts of urban and suburban expansion. Sprawl strains southern Nevada's gray infrastructure of roads, highways, water, sewer, and utilities, it damages the native ecosystems, and it creates sociodemographic inequalities among the population. The problems are not easily resolved; however, today there is a commitment to conservation and green infrastructure that was not always a part of the Las Vegas legacy.

A popular Las Vegas marketing slogan is "What happens in Vegas, stays in Vegas." I was fortunate to work with a great group of people in Las Vegas to complete open space, trail, and green infrastructure plans that have and will continue to benefit the lives of residents, as well as the character and economy of the Las Vegas Valley and southern Nevada, for many years to come. The most important takeaway for me in regard to the work in Las Vegas is . . . if Vegas can do it, any American community can work effectively to conserve valued open space, green infrastructure, and native landscapes that help to define their unique sense of place.

6

Miami Means "Sweet Water"

Miami River Greenway, Miami, Florida

The lower portion of the Miami River Greenway Promenade. Photo by author.

"VIA VERDE?" THE RESIDENT OF LITTLE HAVANA ASKED. "Why is the 'way' green?" she continued. It was clear to me that I was not doing an effective job of communicating the intent of the proposed Miami River Greenway project to my Cuban American audience on a sultry April 2000 evening in Miami, Florida. The word greenway in Spanish, "vias verdes," did not translate well because residents of the neighborhood did not understand why a greenway was important to Miami-Dade County and the project sponsors. Admittedly, I was stumped as to how to convey the benefits of the greenway to neighborhood residents. I needed to better understand the audience for my message and provide them with a frame of reference for what we were trying to accomplish along the edge of their neighborhood in order to illustrate how it would bear relevance to their daily lives.

Why was the community of Miami-Dade County, Florida, interested in a greenway master plan for the Miami River? Several forces were driving interest to this long-neglected urban river. There was a desperate need for removing contaminated sediments from the river to improve water quality, protect the environment, and provide a deeper river channel for increased use by larger boats and commercial shipping. Other interests revolved around saving a unique two-acre parcel of land situated along the edge of the river that contains a sacred, historically significant, archaeological treasure that today is known as the Miami Circle. However, there was more to the story of Miami's efforts to transform the river from a perceived liability to a functional and vibrant urban asset.

Historic Circle

The Miami Circle is a unique archaeological site measuring thirty-eight feet in circumference, with a series of post holes carved into the limestone bedrock. It is the only known evidence of permanent prehistoric settlement in the eastern United States. The land where the circle was discovered is located at the mouth of the Miami River, where the freshwater river empties into the brackish water of Biscayne Bay. The circle is one of the most important historic landscapes of South Florida, estimated to be approximately 2,000 years old.

The circle is now listed on the National Register of Historic Places and is a National Historic Landmark. It was discovered in 1998 during routine investigations required by city ordinance because the proposed development site was within a zone of archaeological significance. The site was slated for condominium development when the team of archaeologists and surveyors, led by Bob Carr, county archaeologist, discovered the unique formation of post holes and recognized that they had stumbled on an incredible historic feature. Almost immediately a community campaign was begun to save the historic landscape from demolition and destruction. The landowner/developer offered to saw cut a large limestone block containing the circle from his project site and move the

Miami Circle at the mouth of the Miami River at Biscayne Bay. Photo by Marc Averette, March 7, 2011.

---

relic to another location in the community. That offer was appropriately rejected by community leaders, which resulted in a January 1999 lawsuit to halt the project. After a few months of frenzied negotiation, the developer was paid $27 million to walk away from his condominium project, with a combination of funds from the State of Florida and Miami-Dade County.[1]

Today, the circle is part of a protected landscape that sits in the midst of one of the fastest urbanizing communities in the Americas and is one of the signature elements of this diverse urban riverfront landscape.

There's a River—in Miami?

I began work on the Miami River Greenway Master Plan in November 1999 as a result of a working relationship that I had been building for many years with the Trust for Public Land (TPL), a national nonprofit organization based in San Francisco, California, that protects land for people. TPL was established in the early 1970s as an offshoot of the Nature Conservancy. It was founded by California board members of the conservancy who wanted to focus their efforts on urban landscapes, where conserving and protecting greenspace is vital to the quality of life. Beginning in 1992, I worked with Will Abberger, a TPL project manager based in the Tallahassee, Florida, regional office, on a variety of greenway projects in North Carolina and Tennessee. TPL was an important member of the team that worked to save the Miami Circle; Brenda Marshall McClymonds, project manager in the Miami field office, led the community efforts to secure funding to buy out the developers' real estate interests. TPL provided a bridge loan of $8.7 million to Miami-Dade County in support of the Miami Circle acquisition. It was Brenda and her coworker, Lavinia Freeman, who reached out to me in June 1999 to discuss the need for the Miami River Greenway Master Plan. The desperate efforts to save the Miami Circle served as a wake-up call to the Miami-Dade land conservation community, who needed a process and strategy for protecting more than just the two acres of land surrounding the circle.

The 5.5-mile study area for the Miami River Greenway begins at the eastern edge of the Miami International Airport and continues downriver to Biscayne Bay. The word Miami means "sweet water"[2] and is derived from the Native American Tequesta tribe, known as the Mayamis, who lived along the riverbanks for more than 1,000 years. As I began work on the Greenway Master Plan, I was surprised to learn that the river was virtually unknown to most Miami-Dade residents—who did not realize that a river flowed through the heart of their community. Even more surprising, few knew that their city was named for the river. Those who did know of the river had disdain for its existence as it was an impediment to crosstown automobile travel and was regarded as a back alley for Caribbean drug trade and other nefarious activities. Not long after my contract was signed with TPL, I conducted a telephone interview with a reporter from the *Miami Herald* who questioned the merits of the greenway master plan. She challenged the vision and goals for the project, saying that the study area was not a safe place for people to walk, bike, or recreate. Subsequently, the *Herald* ran an article about the work we were about to undertake, raising doubts about the wisdom of spending money on a greenway plan. Were it not for the passion and efforts of Brenda McClymonds, Lavinia Freeman, and TPL, the Miami River would have remained a community liability. Brenda was determined to change all that. She needed my help in preparing a master plan, and more important, she needed the master planning process to generate awareness about the potential for the Miami River. The goal was to transform what was perceived by many as a liability into an asset that over time would become a significant urban landscape in South Florida's metropolis.

A Working River

The Miami River has evolved over thousands of years from a brackish tidal channel into a freshwater stream that conveyed the flow of the Everglades into Biscayne Bay. The river is the oldest natural landmark in southeast Florida. Since the 1940s the Miami River has been accurately referred to as a "working river" when, during World War II, it became one of the most important domestic continental harbors for military ship construction and repair.

From 1909 to 1933 the river was lengthened and widened. The Miami Rapids, a natural limestone dam that stretched perpendicularly across the north-south flow of the river, was destroyed in 1909 when the Miami Canal was built as part of the Everglades drainage project. Remnants of the limestone outcrop can be seen today at the City of Miami's Paradise Point Park at NW South River Drive. After dredging for the Miami Canal began, the South Florida water table dropped dramatically and Everglades muck flowed into the once clear waters of the river. Concerns over environmental degradation, water pollution, bridge openings, and the unkempt appearance of the Miami River have been voiced continuously since the 1940s.

Today the Miami River corridor supports a variety of land uses, including heavy and light marine industry, water dependent business operations, restaurants, shops, government facilities, and residential and commercial development. In the late 1990s nearly 20 percent of the nation's Caribbean Basin cargo trade was transported through the Miami River. The river is an international transfer point for shippers, who annually are moving two million tons of cargo to and from nearly 100 ports of call. With an estimated $4 billion in annual cargo volume, the Miami River is Florida's fourth largest port and one of the city's largest employers. During the time frame of the master planning process, riverfront properties were assessed at $1.3 billion and generated approximately $20 million in annual property taxes. Hotels on the Miami River served half a million overnight guests in 2000, representing annual expenditures of

$100 million.[3] Suffice to say, the short and skinny river channel is one of the most unique waterfronts anywhere in the United States, and it would require a very creative planning process to define future greenway development that would be of value to residents and tourists.

Partners in Revitalization

In the mid-1990s TPL began working in partnership with the Miami River Commission (MRC), a specially constituted watchdog and advocate for the Miami River, established in 1997 by the Florida legislature to conduct a comprehensive study for the restoration of the Miami River and Biscayne Bay. Faced with enormous obstacles, including lack of accessibility to the river, neglect of adjacent riverfront properties, crime, illegal drug trade, and other community issues, TPL and the commission worked in partnership on a variety of projects that would transform the river from community liability to regional asset.

In 1999, TPL was tasked by the Miami River Commission to take on the assignment of studying the river corridor and making recommendations for restoration and revitalization. TPL believed that producing the Miami River Greenway Plan would define a greenprint (an action plan for conservation and revitalization) and result in an urban redevelopment strategy that could unlock the potential of the river as a natural resource and public amenity, spurring neighborhood revitalization that would also transform economically depressed areas into more attractive places to live and work. The Miami Circle project became the catalyst that led to the beginning of the greenway planning process.

The Malecon

From the outset of the greenway planning process, one of the most important aspects of my work on the project was to engage the community in the planning process. TPL and the MRC felt it was critical that community residents be included in the design process to build support for transformation of the river corridor. I led two rounds of public meetings in 2000 that included participation in community events, including Miami River Day, where draft versions of the greenway plan maps were displayed as we solicited public input. Public engagement proved to be challenging, as English is virtually a second language in the heart of the Miami community. (According to the 2010 U.S. Census, 70 percent of residents spoke Spanish while 22 percent spoke English as their native language.) Community residents had immigrated to Miami from a variety of Latin American nations, including Cuba, Nicaragua, Brazil, Venezuela, Argentina, Colombia, and Mexico. Communicating the need for a greenway was a challenging proposition as the origins of greenways are derived from English and northern European culture. There are not many parallel landscapes to draw on in Spanish culture, beyond formal private gardens. The most well-known Vias Verdes in Spain makes use of abandoned railway corridors, reconstructed as trails. That is why I was stumped when my Cuban-American audience began to ask why "vias verdes" was a necessary outcome for the Miami River corridor. I then turned to Cuban culture and found a landscape that I hoped would resonate with the Little Havana neighborhood. Along the shoreline of Havana is a promenade known as the Malecon, which is a signature and beloved landscape of the Cuban people. Located along the seawall in Havana, the Malecon offers a broad walkway for strolling, fishing, and other leisure and recreational pursuits. The Malecon is a hot spot for dating, where lovers can sit and watch the waves crash against the seawall. The Malecon is thus affectionately regarded as one of the most important social landscapes in Cuba.

Returning for a second round of public meetings with the Little Havana neighborhood, I proudly proclaimed, "We want to build the Malecon in Miami." This was met with genuine enthusiasm. The woman who posed the original question approached me and said: "You want to build the Malecon along the edge of our neighborhood? Why didn't you say that during the first meeting? We would love to have the Malecon in Miami."

Three Rivers in One

To kick off the master planning process, Brenda and Lavinia invited me to join them on a river cruise so that we could examine the challenges of greenway planning from the vantage of the river. As a result of this examination, I realized that the river was not a homogenous landscape and concluded that for greenway development purposes this very short urban river needed to be subdivided into three different and distinct zones: upper, middle, and lower. Subdividing the river corridor into three zones enabled a greenway design that could more appropriately fit the conditions of each zone. This subdivision proved to be one of the most important aspects of the master plan as it provided the flexibility necessary to propose greenway facilities that could be integrated into the existing urban fabric of the corridor. Additionally, it helped resolve objections to greenway development that were coming from the marine industrial businesses that owned acres of waterfront land in the upper portion of the river corridor, near Miami's international airport. The three different river zones can be described as follows:

Upper River

This extends from the edge of the Miami International Airport to the 22nd Avenue bridge and includes much of the river's marine industrial complex: Miami Industrial park, Grapeland Heights and Melrose residential neighborhoods, the airport, Miami Rapids, Palmer Lake, Curtis Park, and Miami Intermodal Center. The land use is characterized primarily by heavy and light industry, much of which is waterfront or water dependent and serves international cargo and freight needs along with shipbuilding and repair.

With respect to future greenway development along the upper river, it was immediately clear that with so many marine dependent industries located along the edge of the river, there was no opportunity for having waterfront greenway facility development. The existing street grid and other infrastructure that was a block or two off the river were examined as possible locations for greenway trails, conceding that an on-road or roadside solution was the only way of establishing a continuous network of community trails.

Middle River

This extends from the 22nd Avenue Bridge to the 5th Street Bridge and includes the Miami Civic Center, as well as Spring Garden and Grove Park residential neighborhoods and their waterfront homes. The land uses along the river are residential, civic, recreational, and institutional. In the middle river there are numerous constructed inlets that support residential enclaves. Waterfront living is the most prominent feature of the middle river.

With respect to future greenway development along the middle river, as with the upper river, it became apparent that with so many waterfront residences located along the edge of the river, there was no opportunity to promote the idea of waterfront greenway facility development. The existing street grid, neighborhood parks, and other infrastructure that was within a block or two of the river provided corridors for greenway trails.

Lower River

This extends from the 5th Street Bridge to Biscayne Bay at the mouth of the river. This is the most urban section of the river and includes Brickell Key (a separate island), the Central Business District, Brickell, Overtown, and East Little Havana neighborhoods. The key land uses include the Miami Circle, Bayfront Park, Fort Dallas Park, Jose Marti Park, Lummus Park, the Miami Convention Center, Miami Government Center, Miami Riverside Administration Center, and a short section of Miami Riverwalk. At the time of the master planning process, there were some high-rise hotels, offices, and residential buildings.

The lower river was the most likely zone to support river edge greenway facility development. An existing short section of riverwalk was located at the Hyatt Hotel and offered a model landscape to build on throughout the zone. There was also a need to link the riverfront to the nearby Central Business area and Brickell Avenue corridor.

Themes of Greenway Development

In addition to the distinct character zones of the river, the greenway master plan had to address the needs of residents by identifying the most important activities that occur daily throughout the corridor. This was accomplished by describing "themes" of activity that reflected input received from public outreach and engagement, which resulted from meetings with residents and interviews with the marine dependent businesses. The river themes are as follows:

The Working River

As the fourth largest port in the state of Florida, the river is a vital transfer point for international cargo. The marine dependent businesses use the river as a transportation corridor to Biscayne Bay, the Caribbean, and other international ports of call. The river was and is to this day a gritty, blue collar, industrial corridor that on any given day transports enormous ships along its 5.5-mile length. It is an incredible experience to witness a massive freighter being towed up the river for service or repair, and it is what makes the Miami River one of the most unique urban rivers in the world.

Home on the River

The Miami River is home to a large and diverse population. From South Florida's earliest settlement, the river has been a center of residential life, with waterfront homes in styles rustic, historic, classic, and contemporary. Historic neighborhoods rich in tradition and culture include Overtown, Spring Garden, Grove Park, Allapattah, and Little Havana.

The Natural River

The river is an important source of animal and plant life. The endangered manatee, while still at home in the river, is a symbol of the ecosystem at risk. This popular animal—and other species that depend on clean water—are benefiting from efforts to restore the environmental health of the river and surrounding areas. A major dredging project, completed in 2008, removed contaminated sediments from the river and deepened its flow, stabilizing the shoreline and reducing pollutants from the surrounding watershed.

Heritage

The historic and archaeological record of the Miami River tells a fascinating story of human cultural activity, from the earliest Native Americans to nineteenth-century settlers to the courageous immigrants of yesterday and today. The Miami Circle, at the mouth of the river, is a unique artifact believed to reflect the culture of the Native Americans who lived here as long as 2,000 years ago. The archaeological sites reveal glimpses into the lives of the city's more recent settlers. Along the river are also many examples of historic architecture—Fort Dallas, the Scottish Rite Temple, Wagner Homestead House, and the Miami River Inn among others.

The River as a Destination Landscape

The Miami River Greenway offers a fascinating view of the working river, with its daily bustle of shipyards, cargo, and commercial marine activity. The river has become a focal point of tourism and entertainment—a place where visitors and residents gather for dining, shopping, recreation, and cultural enjoyment. With development of the greenway, the Miami River has become a major tourist attraction, blending scenic landscape with lively commerce. Walkways and bike trails, restaurants and cafes, hotels, inns, shopping, and historic sites all combine to make the Miami River an appealing center of activity.

New and Improved Miami River

Combining the character zones and the greenway themes, a series of goals and objectives were defined for the proposed greenway that would provide a flexible development program and which respected the unique characteristics of each zone while at the same time providing a unified approach to facility development along the entire 5.5-mile corridor. These included:

Improve access to the river
Sustain the working river industries and their operations
Restore water quality throughout the river
Serve as a destination landscape for South Florida
Encourage compatible land uses surrounding the river
Foster an ethic of stewardship
Celebrate the multicultural diversity of neighborhoods that surround the river.

As I worked with TPL staff to roll out the plan of action, we noticed that local attitudes about the river began to change. The Miami River

Spring Garden Point Park on the Miami River Greenway. Photo by Darcy Kiefel, courtesy of the Trust for Public Land.

Commission was solidly in support of the vision for a river-oriented greenway as it fulfilled an important mission of their work. Brenda, Lavinia, and I attended more than twenty separate stakeholder meetings, often scheduling four to six meetings in a day. Our message was rather simple: the river is vitally important to the identity of the city, and it is time to invest in this valued natural resource. This message resonated with many of the individuals and groups we spoke with, and soon we began collecting letters of endorsement in support of the greenway plan. A vision for the Miami River was beginning to gain traction across the community. Residents were waking up to the reality that the river was worth the effort of revitalization and transformation. The *Miami Herald* began to run favorable articles in support of the greenway.

Key Greenway Strategies

With the analysis and public engagement complete, and framed by the river zones, community input, and project vision and goals, our team generated a set of recommendations for future development of the Miami River Greenway. The development program consisted of five key elements: (1) improvements and enhancements to river channel banks, (2) improved points of public entry to the river, (3) construction of a primary system of public trails and walkways, (4) improvements and enhancements to existing parks, and (5) improvements and enhancements to existing bridges and roadways.

Improvements to the River Channel

The first action undertaken was to dredge the river, which was completed in 2008 through a unique partnership between the City of Miami, Miami-Dade County, and the U.S. Army Corps of Engineers. Dredging improved navigation and helped to clean the river of mud and debris. Riverbank stabilization restored native trees and vegetation along portions of the river shoreline and created habitat for wildlife.

Points of Public Entry

Access to the Miami River was improved in several ways. First, public parks along the river were reopened and cleaned up and new facilities were added so that residents and visitors could have access to a linked network of trails and recreation facilities. Additionally, neighborhood gateways, along with signage, were added at key locations along the 5.5-mile river corridor. Bridge improvements are being undertaken, improving pedestrian and bicycle access to the river corridor. Most significantly, private land development along the river has opened parts of the lower river to public access and use.

System of Trails, Bikeways, and Walkways

Perhaps the most significant result from a recommendation within the plan was the completion of a comprehensive network of trails, bikeways, and walkways adjacent to the river corridor. The Miami Riverwalk was extended and completed along much of the lower river, on both sides, from Biscayne Bay to the NW 5th Street Bridge. This has improved residents' and visitors' access to an extensive riverfront promenade. West of NW 5th Street, and throughout the rest of the river corridor, a system of on-road bicycle facilities, sidewalk improvements, and off-road trails has been developed that weaves through the existing urban fabric adjacent to the river. Among these improvements is the Overtown Greenway, an urban streetscape corridor from the Winn Dixie Supermarket at NW 12th Avenue and NW 11th Street to Biscayne Bay at Bicentennial Park. The trail system connects neighborhood residents to parks adjacent to the river. Additionally, a system of trails along Biscayne Bay was included in the Greenway Action Plan and has been completed. These trails link Brickell Key to Margaret Pace Park.

Cost of Facility Development

The improvements to the river corridor recommended in the Master Plan continue to be accomplished in phases. Originally it was estimated that the total cost of these improvements would exceed $23 million (2001 dollars). These costs included the physical improvements proposed within the plan, as well as fees for surveying, design and engineering work, environmental permitting, landscaping, and property acquisition. A phase one development work program was prepared to guide initial investment and development.

Sources of Funding

Several sources of public and private funding were identified within the Action Plan. Miami-Dade County, TPL, and project partners have used a variety of different funding sources, often leveraging sources to finance greenway construction. Funding for greenway development has come from the following sources: federal government, State of Florida, Florida Department of Transportation, South Florida Water Management District, Florida Inland Navigation District, City of Miami, Miami-Dade County, Downtown Development Authority, numerous private sector developments, and the Miami Forever Bond.[4]

Operation and Management Strategies

A key element in the success of the Miami River Greenway was improved management and operation of public facilities along the 5.5-mile river corridor. A public opinion survey revealed the preference to have a public-private organization serve as developer and coordinator of the greenway. The Miami River Commission fulfilled the responsibilities of coordinating development of the greenway. The Action Plan also recommended consideration be given to establishing a riverfront corporation that would serve as champion for the plan, but this idea never gained traction. The City of Miami, Miami-Dade County, and the State of Florida contributed to the development and management of the greenway. Private property owners on the river have also played an important role in developing and managing greenway facilities and elements of the riverwalk.

The Greenway Today

The Miami River Greenway Action Plan was accepted by the Trust for Public Land and the Miami River Commission in April 2001. During the past seventeen years, much has been accomplished, and the vision of an urban greenway along the river has been realized. This affords the

Women walking along the lower portion of the Miami River Greenway. Photo by author.

opportunity to compare the recommendations featured in the Master Plan with what has actually taken place in the corridor. The Master Plan described specific recommendations according to river zone. Among the many accomplishments is a comprehensive signage and wayfinding program that has been installed throughout the greenway corridor, using the iconic manatee as a symbol for the river. The greenway also hosts a number of outdoor events and festivals throughout the year in addition to the Miami River Day event. The greenway has become the destination landscape that was envisioned in 2001 when the Master Plan was completed. The following provides a summary of completed work within each zone of the river.

Upper River

The upper river zone was the most difficult in which to imagine substantial change resulting from greenway development. The Master Plan called for creating more bicycle and pedestrian facilities through streetscape improvements and the installation of side paths. The plan recommended improved connections to Miami Rapids Park and Grapeland Park and a greenway trail that was proposed to encircle restored Palmer Lake, the largest water body adjacent to the river.

This river zone has realized the least amount of change since 2001. In 2013, Miami-Dade County completed a design charrette that reimagined the Palmer Lake and upper river zone adjacent to the Miami International Airport. The proposal envisioned a mixed-use development that is transit-oriented, supporting bicycling and walking as major modes of transportation. Palmer Lake itself is envisioned as a destination for high-end hotel and conference facilities. When these redevelopment proposals are realized, the upper river zone will be one of the most transformed areas of the river corridor. Proposed improvements will create a destination landscape that, while substantially different from the lower river, will provide improved functionality for the airport and the adjacent marine industrial landscape.

Middle River

The middle river held promise of near-term change when the Action Plan was adopted in 2001. A proposed extensive network of bicycle and pedestrian improvements along existing roadways would serve to link residential areas to parks. The Master Plan proposed that the city consider a water taxi on the river, along with a "blueway" trail that would offer area residents an opportunity to use the river for both recreation and transportation. The plan also recommended that improvements be made to existing parks: Curtis, Sewell, and Spring Garden Point. Many of these Action Plan recommendations have been achieved. The middle river zone has remained a stable residential area, and connections to local parks and the river have improved the quality of life for residents. The water taxi system is a reality and includes stops at the Hyatt Regency hotel, Epic Hotel, Garcia's, Bayside Marketplace, Miami Beach Marina/ South Beach, Sea Isle Marina/Omni Mall, and other destinations. The water taxi provides convenient service between Miami and Miami Beach at a cost of $15.00 one way and $30.00 round-trip.

Lower River

The lower river of the Miami River Greenway project has undergone the most significant transformation since the plan was completed in 2001. The Action Plan recommended expansion of the riverwalk through the development of a riverfront promenade. Two design details were furnished in the plan to guide greenway development activity.

During the past seventeen years, billions of dollars in private investment has occurred within the lower river zone. This investment has yielded new residential, commercial, and retail development. The Miami River Commission and Miami-Dade County have expanded riverwalk development as part of the private investment, requiring developers to build portions of the riverwalk in conjunction with their activity. Miami-Dade County has also undertaken greenway development through a series of "close the gaps" projects to connect sections of the riverwalk together and form a continuous pathway along both banks of the Miami River. According to Miami-Dade County, as of August 2017, approximately 65 percent, or 6.5 miles, of the originally planned greenway has been completed in the lower river section. Approximately 3.5 miles represent the close-the-gaps projects to be constructed.

Working River Renaissance

In addition to the success of the greenway, the project has also served to support a renaissance of the Miami River as a working river. There was an initial push back during the planning process from the marine industrial community, many of whom believed that a greenway and successful industrial activity were incompatible. At a September 16, 2015, press conference, City of Miami Mayor Tomas Regalado, City Commission Chairman Wilfredo "Willy" Gort, and Horacio Stuart Aguirre, chairman of the Miami River Commission, jointly cut a ribbon on the new Apex Marine facility, located on the river. Mayor Regalado said at the opening: "The Miami River is experiencing the most exciting renaissance in its history. Today, we can call this river the NEW Miami River. Never before has there been such a resurgence in the recreational marine trades industry in Miami. The arrival of Apex Marine to the Miami River solidifies Miami and the Miami River as a major yacht servicing destination. As a long-time promoter of this industry, I welcome Apex Marine as it takes its place along with the many other facilities, such as Merrill-Stevens, Hurricane Cove, Norseman, and others. The opening of this new 5.5-acre facility is indicative of the rebirth of the Miami River as a working river."

Miami River District 2020

The economic success of the Miami River goes well beyond its status as a "working river." This is one of the most unique urban rivers in the world, in one of the most vibrant cities in the world. Brenda Marshall McClymonds and I worked hard throughout the planning process to talk about the importance of the river as an economic engine for Miami and south Florida. At the time it sounded ludicrous to many in the community. Seventeen years later, the river-based economy is flourishing. The Miami River Commission now refers to the area as the Miami River District and anticipates robust growth through 2020. The river has become one of the most desirable real estate markets in South Florida and rivals South Beach as a destination landscape. A December 2016 article in *Forbes Magazine* posed the question: "Why Are the World's Top Real Estate Investors Risking Billions on Miami's Riverfront Renaissance?" That article cited another river greenway project that I worked on, the North Delaware River in Philadelphia, as comparable to the Miami River; both share a challenging industrial heritage and were underappreciated community assets before being rediscovered and developed as a result of greenway plans. The *Forbes* article focused much of its attention on Shahab Karmley, founder of the global real estate firm KAR Properties, headquartered in New York City. Karmley is one of the major contributors to the proposed Miami River District 2020 effort.

The *Forbes* article specifically calls out the $21 million investment in the Miami River Greenway as one of the catalytic projects that has changed South Floridians' thinking about the river, from polluted asset to centerpiece of urban mixed-use development. Karmley is heavily invested in the future of the river and believes in its renaissance. He is proposing a new project in the heart of downtown Miami, called One River Point, that will bring millions of new private sector investment and another shot of energy to the lower river corridor. A quote from the *Forbes* article perfectly sums up the change of attitude in the Miami River: "Instead of cigarette boats packed with cocaine, Karmley envisions luxury motor yachts docked next to up-and-coming Michelin-starred restaurants. Where there were once vacant blocks ringed by chain-link fence and barbed wire, he sees mixed-use retail corridors, waterfront and pedestrian parks, and protected historical sites. Where there was once not a tourist in sight, he anticipates a new generation of Miami travelers flocking to the riverfront first, and Miami Beach afterwards."[5]

From my earliest involvement with greenway planning, design, and development, I could see that one of the most important opportunities for greenways was their ability to provide a significant return on investment. During the past thirty-five years, I have made the case for the economic benefits of greenways in many communities across America and for communities around the world. In my opinion, the greatest barrier to successful greenway development is the commitment to the idea, more so than how much the project will cost. The Miami River Greenway is an excellent testament to this ideal. The City of Miami first had to rediscover and reconnect with its heritage, which occurred with the discovery and preservation of the Miami Circle. Modern Miamians had to overcome their lack of knowledge about the river and its importance to the vibrancy of their community. If the river was a "dead" zone, as some believed, so were other parts of downtown. With the support of the Trust for Public Land, the community overcame the lack of respect for the river. The greenway became a vital part of a reinvigorated and revitalized urban waterfront and fully appreciated for its cultural significance.

Urban Riverfronts Are Public Landscapes

Urban rivers belong to the entire community and should not be restricted to those who can most afford to live on the waterfront. The Miami River Greenway concept has been challenging to execute. Today, not all aspects of the river corridor are open for public access and use. Local government leaders have not always adhered to the recommendations in the Greenway Master Plan, and as a result the riverfront promenade is interrupted by private marinas and private walkways that serve residents of adjacent high-end, multistory residential buildings. This is a violation of the spirit and intent of the greenway, and it further exacerbates issues of exclusion, gentrification, and lack of equitable access. Additionally, Miami-Dade County and the Miami River Commission have accepted and approved proposals for riverfront development that violate the fifty-foot-width envelope established within the plan, at times allowing for as little as an eight-foot-wide corridor for the greenway riverwalk. This also diminishes the experience of using those portions of riverwalk greenway in the lower river section.

When communities limit public access and allow for substandard greenway development to occur along their urban riverfronts, they change the economic values along urban rivers and, more important, the economy of developed landscapes that are located distances from the riverfront. Keeping the riverfront public provides economic benefit that can extend several blocks to a mile from the riverfront. Local governments should resist the temptation to privatize their riverfronts, or diminish the objectives of greenway development standards, in favor of exclusive access and use for private riverside land development.

Jose Marti Park on the Miami River Greenway. Photo by Phil Schermeister, courtesy of the Trust for Public Land.

The Soul of a Community

The most important outcome of the Miami River Greenway was the ability of this unique urban landscape initiative to connect residents to the historic roots and cultural heritage of their community. This is true of many urban greenways, which often follow historic travel corridors, such as abandoned railways, roadways, utility corridors, or rivers and streams. These corridors, or routes of travel, defined the history of founding, settlement, growth, and development. The Miami River Greenway not only connects residents to their cultural heritage but also defines the role that ecology and indigenous people had in shaping community settlement and development patterns. Today, as people walk along the greenway and are linked with the Bayshore Trail, they are provided the opportunity to learn about the history of Miami from its humble beginnings to modern times, as one of America's most fascinating and dynamic urban centers.

Unintended Consequences

Gentrification can be one of the most significant unintended consequences of urban greenways. Several high-profile urban greenways, including the High Line in New York City, the Atlanta Beltline, and the 606 in Chicago, have come under warranted scrutiny for the unintended negative impacts they have had on surrounding minority and economically challenged residential neighborhoods. The Miami River Greenway shares this unfortunate consequential history.

Many proponents of urban greenways rightfully tout that they will increase access to greenspace for adjacent neighborhoods and that they will improve the economy in part by increasing the value of homes in those neighborhoods. The paradox is that successful greenways may also increase the economic value of adjacent housing stock, pricing out of the marketplace the very people that they were intended to benefit. This has become one of the unintended consequences of the Miami River Greenway. As one example, the Little Havana neighborhood, which borders the greenway, has been negatively impacted by gentrification due to dramatically higher housing costs. In simple terms, higher rents, higher taxes due to improved home valuations, and higher sale prices for new and existing housing have become a problem for existing Little Havana residents, resulting in gentrification. Part of the increase in housing costs is due in part to the success of the greenway; however, there are many other economic factors at work in Miami that have led to higher housing prices in the downtown area. A portion of this change in economic value occurred as a result of the greenway transforming what was once a neglected liability into a successful urban amenity. Greenway development made the riverfront and lands adjacent to the river corridor more attractive for investment, development, and redevelopment. As a result, the cost of rents and sale prices for homes in Little Havana increased.

The Miami River Greenway alone cannot and should not shoulder the entire blame for changes in the affordability of Little Havana residential homes and other urban neighborhoods surrounding the river. The impact is complex and must account for the dynamics of living in the heart of modern American cities. The solutions to gentrification are also complex and not easily resolved. They involve a variety of targeted economic solutions, including a concerted effort to maintain affordable housing, instituting rent control, providing financial incentives to local small businesses so that they can remain within the affected neighborhoods, and most important, constant communication with residents.

7

Lowcountry Life

Charleston County Greenbelt Plan, Charleston County, South Carolina

The Lowcountry of Charleston County, South Carolina, contains vast expanses of marshland that border broad brackish water creeks and rivers. Photo by author.

DICKIE SCHWEERS HAS BEEN A RESIDENT of the Charleston County Lowcountry all his life, following in the footsteps of generations of his family. As we rambled along the dirt roads of the Francis Marion National Forest in his pickup truck on a hot August 2005 afternoon, he concisely articulated the lifestyle of a Lowcountry resident and why it is so important to protect their way of life: "We have all grown up on the water. We fish and crab just like our parents and grandparents did. Our homes are built in the marsh. Charlestonians have a strong, emotional connection to this land. We are as rooted here as the longleaf pine and palmetto palm. To lose this land and its resources is to lose our way of life. That is why it is so important to me, and to so many Charlestonians, that we act now to protect this land and lifestyle for ourselves, our children, and grandchildren." Dickie's response encapsulated both the need and support for the Charleston County Greenbelt Plan, the first effort by Charlestonians, in the more than 400-year history of their community, to define a comprehensive vision for land and water conservation.

The residents of Charleston County backed this vision with a twenty-five-year sales tax program in a November 2004 referendum that generated $220 million in financial support for land and water conservation. Dickie Schweers, who served as a member of the Charleston County Council (2006–2018), along with other members of the County Council, a specially appointed Greenbelt Advisory Board, and Charleston County staff, created one of America's most successful greenspace conservation programs, which protected more than 38,000 acres of land between 2006 and 2017.

Lowcountry Defined

The Lowcountry of South Carolina is both a geographical and a cultural landscape and as such is one of the most unique and distinctive landscapes in North America. The city of Charleston is regarded as the unofficial capital of the Lowcountry. Beyond its aesthetic and lifestyle appeal, the Lowcountry provides many important benefits. It encompasses thousands of acres of salt marsh estuary and riverine ecosystems, barrier islands, and forested freshwater wetlands, such as those we were touring with Dickie in the Francis Marion Forest. This combined system is important for carbon sequestration, serving to absorb large amounts of carbon dioxide and transforming it into energy for continued growth of the ecosystem. Research has shown that forests across the planet store approximately 45 percent of terrestrial carbon and can sequester more than 2.6 billion metric tons of carbon dioxide (CO_2).[1] In Charleston County alone, a study concluded that the current forest system sequesters approximately 13,500 metric tons of CO_2.[2] Carbon sequestration is becoming increasingly important as the continued release of CO_2 and other greenhouse gases are contributing to global warming and global climate change.

Few communities in the world are more susceptible to changes in global climate than Charleston County, due to its elevation above sea level and proximity to the Atlantic Ocean. When the Lowcountry vegetation and topsoil is stripped away and transformed into houses, roads, businesses, and other urban uses, not only do residents of Charleston County lose the benefit of natural carbon sequestration, but they also expose themselves to increased risks associated with tidal flooding and saltwater intrusion and, as Dickie succinctly stated, a diminished quality of life. Tidal flooding and more severe storm events, in the form of hurricanes and "noreasters," are an annual threat to life in Charleston County. The community has survived numerous direct hits from major hurricanes. Hurricane Hugo in 1989 was one of the most devastating to impact the community. Thus, the salt marsh and barrier island system is crucial to the continued survival of the community. Not only do county residents depend on groundwater for drinking water but also for retaining proper baseline flows in local freshwater streams, rivers, and marshes, which serves to maintain the delicate balance that supports all ecosystem life. Fishing and crabbing in the Lowcountry is made possible by the mainland river and stream freshwater flow into the saltwater marshes along the tidal shoreline.

The Lowcountry is more than just a beautiful place to live; it is most importantly a vital ecosystem that supports the type of human life and activity that has attracted and sustained generations of Charlestonians. The citizens of Charleston County are intertwined with the unique landscapes and ecosystems of the region. To carelessly destroy this landscape through pervasive land development practices has severely impacted the land and water that support the Lowcountry lifestyle. Dickie Schweers was correct in his assessment of the need for a countywide conservation plan, and his vision and understanding were shared and supported by thousands of Charlestonians.

Funding Greenspace Conservation

The county did not begin its Lowcountry conservation effort working from an approved plan; it began with a vote, and not just one vote, but several ballot box attempts to approve appropriate referendum language. Charleston County strongly believed that a sales tax, sometimes referred to as a consumption tax, was the best approach to meet community needs for both transportation and land conservation. With tourism being one of the largest economic engines in the community, a sales tax was viewed as the best option for collecting funds needed to meet program goals, as there was a direct correlation between tax collection, expenditures of those funds, and public benefit. Residents and tourists were using local roads and highways in record numbers, and the impact of this use resulted in congestion and degradation of roadway conditions. Residents and tourists also use the Lowcountry for a variety of outdoor pursuits and purposes, and yet those landscapes were disappearing at an alarming rate. Charleston County quantified some of this need, stating that the county has:

1,400 miles of state, county, municipal, and community roads with gridlock imminent and projected roadway improvements of more than $1 billion;

mass transit needs that greatly exceed revenues, totaling $335 million in operating and $100 million in capital costs;

commissioned a "Greenbelt" referendum study committee, which identified 46,100 acres of strategic lands needed for conservation, at a cost of $122 million; and

recently completed the Cooper River (a.k.a. Ravenel) Bridge at a cost to the local community of $75 million in matching funds, and the county, with no funding available, had to make its first bridge payment by January 2004.

The first attempt to enact a sales tax for transportation improvements and greenspace conservation occurred in 2000. The referendum language proposed a twenty-five-year sales tax program with proceeds split 45 percent for road construction, 30 percent for mass transit, and 25 percent for greenspace. The referendum failed by 932 votes, with reasons for the defeat listed as lack of a detailed spending plan, too short a time frame to fully consider the merits of the referendum, lack of outreach and education about the sales tax, and lack of a unified community approach.

A second attempt at the sales tax was made in 2002 and included more definitive plans from each community within Charleston County for meeting transportation needs, a unified effort by the entire community, and a comprehensive survey of citizens. The allocation formula changed to 65 percent for road construction, 18 percent for mass transit, and 17 percent for greenspace. This time, public outreach was led by the Charleston County Chamber of Commerce and local municipal leaders. An antitax opposition group was launched to defeat the referendum. The public was confused about the spending plan, and the referendum failed by 685 votes. This time, the reasons for defeat included not enough detailed spending information, wording of the ballot language that was considered biased, and a lack of appropriate advocacy for the sales tax.

Shortly after the 2002 referendum, the county was sued by an antitax lobby regarding the ballot language, and the South Carolina Supreme Court voided the election results in August 2003. Charleston County requested the opportunity to hold a special election in January 2004, but the governor denied the request and at the same time signed an Executive Order putting the referendum on the November 2004 ballot. Meanwhile, with the clock ticking on financial commitments, the county scraped together enough money to make two payments on the Cooper River Bridge. The mass transit agency, CARTA, ran out of money and cut its transit routes by 75 percent to remain operational.

The third referendum occurred in November 2004. The county changed the ballot language to meet legislative and South Carolina Supreme Court requirements. A detailed spending plan was developed, and the sales tax program made a provision for public input, annual audits, plan review, and citizen committees. The Chamber of Commerce led efforts to educate citizens about the referendum. A coordinated marketing and advertising campaign ensued two weeks before the fall election. This time, the referendum was approved by the voters, with 58.8 percent in favor to 41.2 percent opposed. The measure was approved by 19,110 votes. The success of the third referendum was tied to the relationship formed between local governments, the business community, and other community organizations. *Charleston Regional Business Journal* publisher Bill Settlemeyer attributed success to the impact that increased traffic congestion had on voters and on voters' recognition of the need to preserve greenspace for future generations of Charlestonians.

The Trust for Public Land (TPL), a San Francisco–based nonprofit land conservation organization, completed a summary of similar ballot measures approved between 1988 and 2010. TPL's "LandVote" report noted that of the 2,299 ballot measures put before American voters during that timeframe, 1,740 of these measures were approved with $56 billion in funds appropriated for land and water conservation purposes. TPL concluded

> that in robust and challenging economic times alike, American voters have strongly supported conservation finance measures that preserve natural lands, create parks, and protect farmland. Communities that have approved conservation measures are as widespread and diverse as the purposes to which they have committed funds. Conservation finance ballot measures have passed in urban and rural communities in nearly all 50 states. Funds approved by individual measures range from a few hundred thousand dollars to several billion dollars and have supported purposes including parks and playgrounds, farmland preservation, watershed protection, trails and greenways, forests, and wildlife habitat.

The November 2004 Charleston County Sales and Use Tax was enacted for the protection and promotion of "the health, safety, welfare, and quality of life" of the Charleston County residents by financing the costs of highways, roads, streets, bridges, and transportation-related improvements along with investments in the mass transit system and greenspace conservation (a.k.a. Greenbelts). The referendum defined the maximum funds to be collected and spent on Greenbelt projects during the twenty-five-year life of the sales tax as $221,571,200. Charleston County formally began collecting the Transportation Sales Tax on May 1, 2005.

"We Need a Plan"

My involvement with the Greenbelt Plan began with a March 2005 telephone call from Mark White, a partner in the land use law firm White and Smith, LLC, based in Kansas City, Missouri. Mark served as a consultant on the Charleston County Comprehensive Plan and was attending the American Planning Association national convention in San Francisco. He was speaking with Charleston County planning director Jennifer Miller about crafting the Greenbelt Master Plan to satisfy the requirements of the approved sales tax. Although we had never met or worked together before, Mark was familiar with my open space planning work in other communities and recommended me to Director Miller.

I immediately arranged a follow-up phone call with Jennifer Miller and Dan Pennick, assistant planning director, and we discussed the specific needs of Charleston County. A month later, the county issued a request for proposals, an invitation to submit qualifications, scope of work, and associated costs for assisting the county in preparing the Greenbelt Plan. The county needed a plan and program to match the language articulated in the voter-approved referendum. My firm, Greenways Incorporated, was shortlisted by the county for an in-person interview, which I attended, and was subsequently hired to complete the plan. During the contract award, Director Miller clarified that Greenways Incorporated needed to serve in the capacity of county staff for the project as the county did not have dedicated staff for the planning effort. That changed several months into the job when the county hired Ms. Cathy Ruff as project officer for the plan.

At the time of my selection as Greenbelt Plan consultant, Charleston County was also in the process of establishing a Greenbelts Advisory Board (GAB). The GAB comprised fourteen citizens and was intended to be the "primary County Council advisory body for Greenbelts." Included in the purpose of the Greenbelts Advisory Board was the charge to develop a Comprehensive Greenbelt Plan for presentation and approval by the Charleston County Park and Recreation Commission and Charleston County Council. The board was chaired by Ms. Louise Maybank, a highly regarded Charleston County resident who was a past recipient of South Carolina's highest civilian award, the Order of the Palmetto, for her work in conservation. Ms. Maybank was also founding director and a twenty-five-year member of the Lowcountry Open Land Trust, which had worked effectively to conserve Lowcountry lands and waters throughout the region. Jennifer Miller and Dan Pennick arranged for me to meet with Ms. Maybank prior to the scheduled meeting with the entire Greenbelt Advisory Board. Ms. Maybank was an effective leader of the GAB as she was visionary, level-headed, and respected by every member of the board. She also had the respect of County Council, staff, and the community at large.

The GAB was a politically appointed group, and not everyone seated at the table was a proponent of the Greenbelt Plan. Of great interest to me was the fact that there was lack of agreement on what the GAB was tasked to accomplish. Further, there was no unified functional definition of the word "greenbelt." The GAB was provided its marching orders by the Charleston County Council and given ten months to produce a plan that satisfied the intent of the voter referendum. As we all were starting from ground zero, I felt my first task was to offer the GAB background information and educate the members about different types of open space and greenspace that should be considered for stewardship and conservation. The GAB had other ideas. In what would become a repeated question at every meeting, the members wanted me to explain how the county was going to divide and spend the sales tax funds. My reply to this frequently asked question was: "The Plan that you produce will determine the funding formula and distribution."

Cities and Villages Surrounded by Green

The first objective in crafting the Greenbelt Plan was to establish a firm foundation for the plan that would support sales tax collection and expenditure. It was fundamentally important for the GAB to build a defensible argument for every element of the program and every decision that they authored for the plan. I anticipated that there might be legal challenges to the plan, and therefore the decisions and actions of the GAB needed to be based on science and South Carolina land use law. At the same time, the plan needed to be visionary and embrace simple and clear reasons why Charleston County should be undertaking it. It was not easy to accomplish, but eventually the GAB settled on a vision for the plan, which was titled "Cities and Villages Surrounded by Green." This began with the crafting of a working definition for Greenbelts, which was as follows:

The term greenbelts will be used to describe a variety of lands that may include public or private lands in rural, suburban and urban settings, including passive greenspace, active greenspace, low-country ecosystems, productive landscapes, heritage landscapes, corridors, natural (green) infrastructure and reclaimed greens-pace. These greenbelt resources collectively form a protected living system of landscapes that serve the residents, businesses, visitors and future generations of Charleston County by preserving and improving the quality of life for all.

Two important actions were undertaken at the beginning of the planning process. My team at Greenways Incorporated completed an inventory of existing greenspace resources, revealing that of the roughly 670,000 acres that comprised Charleston County, approximately 160,000 acres were already classified as protected and conserved greenspace. We prepared a map that illustrated where this existing greenspace was located within the county. The map illustrated an important point about the lack of equitable distribution of greenspace. Without the benefit of a plan, over the course of many years, greenspace had been conserved, preserved, protected, or acquired by Charleston County and its partners in a haphazard manner. The result was an uneven and inequitable distribution of greenspace across the county. Therefore, one of the goals of the Greenbelt Plan would be to focus on a program of action that would address and resolve this issue.

A second important action was to engage the citizens of Charleston County in the planning process. This was accomplished by conducting open house–style public input workshops across the county, conducting one-on-one meetings with key stakeholders, publishing a Greenbelts newsletter that was distributed in both printed and electronic format, crafting a Greenspace Glossary of terms that defined and illustrated the different types of greenspace featured in the working definition, working with local media to conduct feature articles and interviews, and participating as a guest speaker at local events. The public was also invited to access both online (Internet-based) maps and comment forms to gather additional input. The Greenbelt Plan public engagement effort paid dividends as county residents felt that they were provided with a variety of options for being heard. The county staff and GAB were satisfied that the public engagement strategy was working, and it was helping to guide the Greenbelt Plan recommendations.

The "Cities and Villages Surrounded by Green" was eventually transformed from a simple statement into a program of action, which became the foundation of the Greenbelt Plan. Specifically, the vision statement, public input, and detailed evaluation of community need created the following action program:

Rural preservation of land was paramount, and the county needed to act quickly to conserve the land before it was lost to development.

Public access to water was critically important as it was under threat of permanent loss.

Likewise, public access to a variety of greenspace, including parks and trails, was needed.

There was a need to focus on the reclamation of brownfields and other urban landscapes.

The citizens wanted to link greenspace with a countywide network of trails and bike paths.

The ability to leverage the county's sales tax funds with other federal, state, and private sector funds was critically important.

The plan should address the needs of both rural and urban areas.

The vision statement, "Cities and Villages Surrounded by Green," was depicted as a graphic illustration to accomplish this goal and provided a basis for the next steps in the planning process. The map was paired with a simple illustration of the connection among all elements of the Greenbelt system, known as the hub-and-spoke concept. Taken together, these conceptual illustrations helped communicate the intent of the Greenbelt Plan.

Don't "Take" My Land

My prediction that our greenbelt planning work would be met with legal opposition occurred early in the process when I received a phone call from a local landowner who was threatening to sue both me and the county. The basis of his complaint was related to his fear that the county was eventually going to take his land through eminent domain (sometimes referred to as condemnation), otherwise known as local governments' use of regulatory authority to force landowners to sell their land for public use, with compensation. In anticipation of this he was prepared to preemptively sue to stop the Greenbelt Plan before it had a chance to become part of the County's Comprehensive Plan. He used intimidating language during our phone conversation and said he was including me as a part of his proposed lawsuit as I was the consultant hired by the county to help craft the plan. My response was first and foremost that I did not

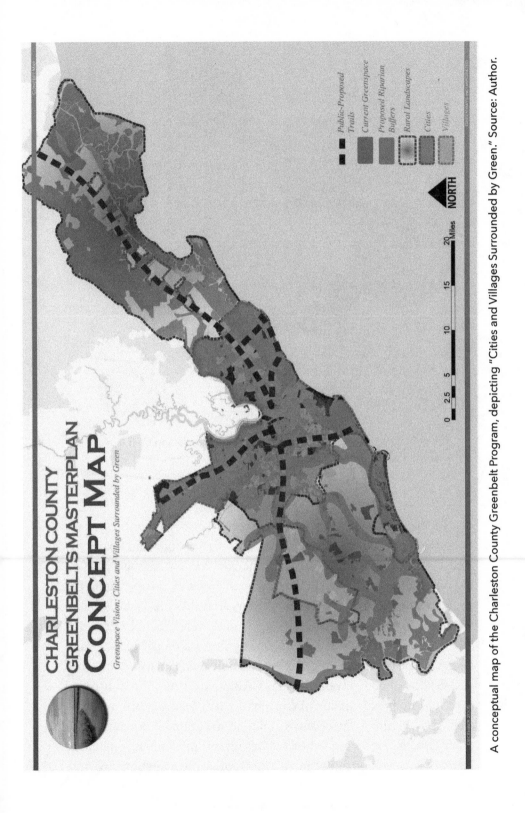

A conceptual map of the Charleston County Greenbelt Program, depicting "Cities and Villages Surrounded by Green." Source: Author.

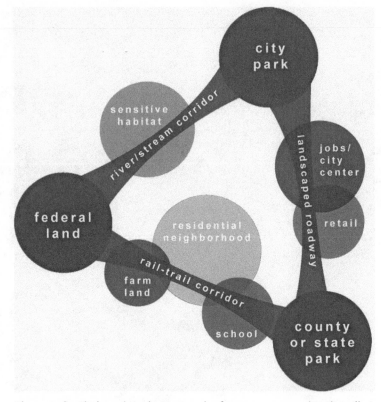

The popular "hub-and-spoke" network of green space and trails is illustrated in this diagram. Source: Author.

take kindly to threats of lawsuits, and second, the open space conservation program would be based on voluntary participation and not on the forceful taking of land. I concluded the conversation saying that if he did not want to participate in the Greenbelt program, no one within the county was going to force him to do so against his will. Likewise, if he wanted to participate in the program voluntarily, he would be provided with a range of options that might benefit him as a property owner and taxpayer. The phone call ended more amicably than it began.

After that phone conversation, the GAB, at one of its bimonthly meetings, took up the discussion of eminent domain. Several GAB members wanted to eliminate the county's ability to use eminent domain, while other GAB members argued that it was unwise to eliminate one of the county's conservation tools from the local government toolbox. The GAB eventually agreed that it was important for the citizens of Charleston County to know and understand that the Greenbelt program would indeed be based on voluntary participation. There would be no attempt to condemn land from unwilling landowners as part of the plan. This strategy worked well in Charleston County as it helped neutralize opposition to the plan. The actions of the GAB were shared with citizens through newspaper feature stories, radio interviews, and on the nightly television news, and removing the threat of condemnation helped calm the opposition and clarified the vision, mission, and goals of the Greenbelt program.

How Much Open Space Is Enough?

Another important issue that had to be resolved in the Greenbelt Plan was determining an appropriate level of service for open space conservation. Based on the inventory of existing open space, roughly 160,000 acres of land had been protected and conserved across the county. The question to resolve was "How much open space should be conserved?" If the GAB was going to define a proper balance between developed land and undeveloped land, it needed to address issues of quantity, location, and distribution. To answer these questions, we combined art, science, and demography to recommend an appropriate level of service for the Greenbelt Plan.

To accomplish this, we examined national standards that described how much open space should be provided for community residents, such as those established by the National Recreation and Park Association (NRPA) in Washington, DC, which specifies a ratio of 6 to 10 acres of parkland per 1,000 residents. However, the NRPA standards did not account for green infrastructure need, natural ecosystem preservation, or climate change. We looked at a variety of other measurements and wound up including the following factors in the decision-making process:

NRPA Standards
Charleston County–adopted policies
County demographic trends
Leisure activity trends from the Charleston County Parks and Recreation Commission

South Carolina Comprehensive Outdoors Plan

County Comprehensive Plan

Opinion surveys on the subject

Statistical surveys on the availability of parks and natural resources lands

Other public input

Market reckoning (availability of land and cost of acquisition)

Benefits of greenspace

Community comparison with other counties that had completed open space plans

Since the sales tax was a forward-thinking program extending to the year 2029, it was important that the Greenbelt Plan forecast future growth and future land conservation strategies. Additionally, the inventory work revealed that most of the existing greenspace had been conserved and protected by federal and state actions, not local initiatives. Approximately 6,700 acres of the total 160,000 of protected and conserved open space was the result of local government action. The County Park and Recreation Commission concluded that future regional park needs alone defined a need for 5,160 acres of new parkland—a 100 percent increase in local government–sponsored greenspace.

We also compared Charleston County with other open space conservation efforts in Wake County, North Carolina; Gwinnett County, Georgia; and Jefferson County, Colorado, to provide a baseline of measurement for the Greenbelt Plan. We shared with the GAB national trends and examples of open space conservation in other communities, citing the Woodlands community in Houston, Texas, and its 30 percent set aside for greenspace; New York City's 30 percent open space goal; San Francisco's desire to protect 25 percent of its land area as greenspace; and Summerlin in Las Vegas, Nevada, a Howard Hughes Corporation planned community that set aside 30 percent of its land area as greenspace.

I proposed for GAB consideration a philosophical approach for establishing an open space protection, conservation, and preservation goal. What if the county were to preserve, conserve, and protect 20 percent, 30 percent, or 40 percent of the total land and water as open space? A simple chart illustrated the philosophical differences and proved to be an effective way of communicating the value of establishing a goal or target

Percentage Goals for Open Space Protection

Land Category to Be Protected	20 Percent	30 Percent	40 Percent
Flood prone land protected	Yes	Yes	Yes
Water quality, recreation, and wildlife lands	No	Yes	Yes
Conservation of land through development (conservation subdivision)	No	Yes	Yes
State and federal forest and game lands	No	No	Yes
Narrative description	Protects all flood prone lands but does little else in terms of conservation	Protects flood prone lands; targets ecologically sensitive lands, greenway corridors, and provides for parkland and prime farmland	Does the same as 30 percent goal; adds state and federal game lands, forests and parks

for open space conservation. It also answered the level of service question. After much discussion, the GAB agreed that a 30 percent target was an appropriate level of service, meaning that the county should strive to protect approximately 200,000 acres of its land and water area as greenspace. This provided the GAB with a numerical target of 40,000 acres of land (200,000-acre target, less 160,000 acres already protected) as the greenspace preservation goal for the Greenbelt Plan.

Greenbelt Lands

Establishing the numerical target was a significant achievement for the Greenbelt Plan. It opened a line of discussion for how the intent of the voters could be transformed into a plan of action. With 40,000 acres of land and water conservation as the goal, the GAB proceeded to define a strategy to achieve the goal. One way to achieve the goal would be to divide the responsibility of conservation and protection among different stakeholders. It was a given that the County Park and Recreation Commission was going to satisfy their need for at least 5,000 acres of land. Charleston County was blessed to have a very strong and diverse private sector conservation community that had a history of achievement. Ms. Maybank, as chair of the GAB, was a leader in that effort. The State of South Carolina and the federal government also had expressed needs of conservation to fulfill. I proposed the following strategy for consideration.

Charleston County Conservation Land Division

Charleston County	8,000 acres
Charleston County Park and Recreation Commission	8,000 acres
Private Sector Conservation	15,000 acres
State of South Carolina	5,000 acres
Federal Agencies	4,000 acres
Total	40,000 acres

This, if nothing else, demonstrated how the 40,000-acre goal could be achieved by spreading the responsibility among different project partners. After weeks of discussion and consideration with the GAB, we recast the land conservation goal according to geographic area rather than institutional need. The final subdivision was as follows.

Division of Land Conservation

Rural Greenbelt Lands	16,204 acres—41 percent of total goal
Francis Marion Forest	10,275 acres—26 percent of total goal
Lowcountry Wetlands	5,610 acres—14 percent of total goal
PRC Regional Parks	4,675 acres—11 percent of total goal
Urban Greenbelt Lands	2,000 acres—5 percent of total goal
Greenway Corridors	1,200 acres—3 percent of total goal
Total Goal	40,000 acres

Each of the Greenbelt component landscapes included a detailed description of what would be achieved and conserved. Geographic information system (GIS) maps visually depicted the locations of each landscape that matched the written descriptions. For Rural Greenbelt Lands the goal was to define a wide variety of landscapes, from agricultural to private landholdings, that were likely to be developed in the future. The Francis Marion Forest is the largest preserved landscape in the region, and the goal of the plan was to purchase private infill lots. Lowcountry Wetlands are critically important to both the ecosystems and built environment of Charleston County. The Charleston County Park and Recreation Commission's need for 5,000 acres of new parkland was included within the November 2004 ballot with the goal of building a network of new regional parks to meet the needs of the county's growing population. The goal of the Urban Greenbelt Lands category was to meet other urban park needs of the municipalities. The final category, Greenway Corridors, was in response to the most frequently expressed need as defined by our public outreach and engagement. Charleston County residents wanted more options for bicycling and walking, and many felt that the best way to achieve this was through a comprehensive network of greenways. The plan recommended a 200-mile network of trails throughout the county.

An overall Greenbelt Plan map illustrated how the individual component landscapes would be interconnected. These maps were helpful in conveying the breadth and scope of the Greenbelt program and how it would serve the needs and interests of all county residents. The GAB agreed that the visual representation of the plan met the voter intent as defined in the November 2004 referendum.

A Plan of Action

The Greenbelt Plan also needed a framework for action. GAB members were continuing to ask me about dividing up and appropriating the sales tax funds. I felt that there were four clear actions for the Greenbelt Plan to resolve:

Allocating the sales tax proceeds
Using criteria to prioritize the expenditure of those funds
Providing the county with a set of tools to achieve conservation objectives
Determining an efficient implementation structure to carry out the plan objectives

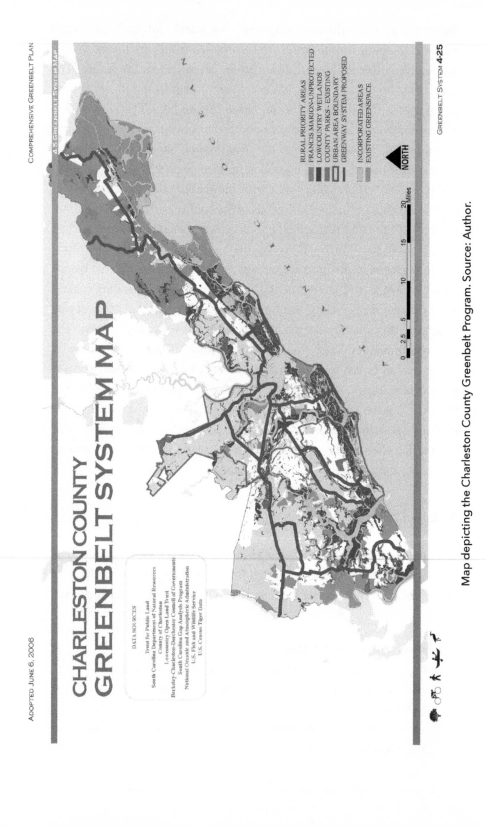

4.5 GREENBELT SYSTEM MAP

CHARLESTON COUNTY
GREENBELT SYSTEM MAP

DATA SOURCES

Trust for Public Land
South Carolina Department of Natural Resources
County of Charleston
Lowcountry Open Land Trust
Berkeley-Charleston-Dorchester Council of Governments
South Carolina Gap Analysis Program
National Oceanic and Atmospheric Administration
U.S. Fish and Wildlife Service
U.S. Census Tiger Data

RURAL PRIORITY AREAS
FRANCIS MARION-UNPROTECTED
LOWCOUNTRY WETLANDS
COUNTY PARKS - EXISTING
URBAN AREA BOUNDARY
GREENWAY SYSTEM PROPOSED

INCORPORATED AREAS
EXISTING GREENSPACE

NORTH

0   2.5   5   10   15   20
                                  Miles

GREENBELT SYSTEM 4-25

Map depicting the Charleston County Greenbelt Program. Source: Author.

The most difficult of the four subjects was allocating the sales tax proceeds. When I signed the contract with the county I was not aware that my recommendation on how to spend the funds would be considered. As the planning process continued, I kept deferring the answer to the question until it became apparent that the county wanted me to provide a recommendation on how the sales tax funds would be divided and spent.

I believed, from the outset of the planning process, that the intent of the voters was influenced by a desire to balance land development with appropriate levels of conservation. As I thought back to the initial conversation with Dickie Schweers, and as our public engagement process helped determine, Charleston County residents wanted to protect, preserve, and conserve their native landscape before it was lost forever to development. This greatly influenced my thinking and recommendations when it came to how the sales tax funds should be allocated.

I offered the GAB different formulas of sales tax allocation for discussion and consideration. One formula I proposed was to allocate the sales tax funds according to population, in which case 91 percent of the funds would have been devoted to the urban areas and 9 percent to rural areas. A second method was to allocate according to geography, with 24 percent of the funds allocated to urban areas and 76 percent allocated to rural lands. A third method was to allocate the sales tax in accordance with where it was collected, with 75 percent of the collections coming from urban areas and 25 percent coming from rural areas. A fourth method was to allocate by registered voter, where 75 percent of all registered voters lived in the urban areas and 25 percent lived in rural areas.

My final recommended sales tax allocation was based on a number of factors—what I felt was the intent of the ballot language, the working definition that had been crafted by the GAB, the results of the public engagement process, and the challenges of land conservation that would be faced during the next twenty-five years. I suggested that 70 percent, or roughly $154 million of the funding, should be dedicated to rural lands conservation and 30 percent, or roughly $66 million of the funding, to urban areas greenspace need. My recommendation was not well received by the mayors of the populated urban areas, who correctly argued that their voters were the ones who made the sales tax possible. My response was that this was a countywide program, and the quality of life for a citizen of any given municipality was greatly influenced by the availability of protected and conserved open space across the entire county. I also believed that there were more opportunities to protect valued open space at affordable prices in the rural areas of Charleston County than in the urbanized landscapes. My final thought was if we skewed distribution of the funds toward urban needs, we would wind up with a very expensive urban parks program that I did not believe was the intent of the voters. To my surprise, Charleston County officials backed me on the division and distribution of sales tax funding.

Raising the Borrowing Cap

Resolving the sales tax allocation and distribution was a critically important step in the planning process. I was troubled by one other important fact regarding the sales tax, and I decided to broach the subject with both the GAB and county staff. The sales tax program was set up to collect small amounts of funding each year for twenty-five years. When I asked the county staff to estimate how much money would be collected in a given year, it varied from around $5 million to $10 million per year. That was a problem as far as I was concerned, as the rate of collection did not provide the county with enough funding to fully meet the needs of the program. There was no way that such small amounts of funding could compete with the explosive urban growth that was occurring in the county. The entire program was at risk of failing due to inadequate levels of funding.

I recommended that the county bond the sales tax and make every effort to raise a large amount of funding immediately to implement the objectives of the program. To bond the program would involve issuing General Obligation (GO) bonds, which are distinct from other types of local government bonds because they were backed financially by the voter-approved sales tax. County attorney Joe Dawson and bond attorney Charlton de Saussure explained to the GAB how the issuance of bonds would work to fund the Greenbelt program. They proposed several different scenarios for issuing bonds, including having the individual municipalities issue their own bonds. One of the most important concerns that they raised during their presentation was the fact that the county could not issue any additional GO bonds, as this would exceed a state of South Carolina–mandated 8 percent borrowing cap (the county was already at the upper limit). To raise the ceiling would require voter approval. I suggested that Charleston County ask the voters for approval. The GAB agreed as did the county staff and County Council, and subsequently a referendum to raise the borrowing cap was put to a vote in November 2006 and was approved by the voters. This again was significant as it eliminated another hurdle for the program, which was to capture the benefit of the sales tax at the beginning of the Greenbelt program and raise enough money at the outset to implement the recommendations of the plan.

When I concluded discussions with the GAB concerning the issue of bonding the sales tax, I clarified that, even with $220 million in funding, this would not be enough to meet the objectives outlined in the plan. I suggested that the county needed approximately $1 billion in funding to accomplish the Greenbelt program goals. To achieve this, the county would need to leverage at a ratio of $4 or $5 of partner funding for every $1 of sales tax funds spent. My recommendation surprised members of the GAB who believed that the bonded sales tax funds would be sufficient to meet the plan objectives. However, they quickly came to realize that in the competitive Charleston County real estate marketplace, my recommendation made sense.

Conservation Toolbox

The complexity of satisfying the intent of the voters was becoming clearer to the GAB the more we worked our way through the implementation strategies of the Greenbelt program. We addressed and resolved many complex issues, transforming the referendum language into a plan of action. Most important for the GAB, they had resolved how the sales tax funds were going to be collected and spent to achieve the plan objectives. I felt that they needed to take a broader look at land conservation, beyond the expenditure of sales tax funds, and consider the full range of tools at their disposal.

I introduced for GAB consideration and discussion a "conservation toolbox." My message concerning the functionality of the toolbox was a simple analogy: if a screwdriver is needed to complete the conservation objective, then reach for the screwdriver, not the hammer or the pliers. In other words, Charleston County needed a full range of tools to maximize both the effectiveness and efficiency of the Greenbelt program. Working with White and Smith, LLC, who specialized in land use law, we were able to identify and include thirty-two different tools that would support land conservation. The toolbox was divided into four broad categories: regulatory tools, acquisition tools, donor or gifting tools, and management tools. Each tool was described, and the benefits and drawbacks of each tool were also defined. I explained that tools could be used in combination to achieve conservation objectives.

Grants Program

The only outstanding issue to resolve was how Charleston County was going to implement the recommendations in the Greenbelt Plan. The agency charged with crafting the plan, Charleston County's planning department, was not staffed to take on this challenge. There was no existing department or agency that was prepared to take on the role of champion and implementer of the plan. I worked with the GAB to craft a framework for implementation.

With the division of funding defined, it was decided that the county would operate a grants program as a way of distributing bonded sales tax funds to priority land conservation projects. A rural lands grants program would be operated by a newly formed Charleston County Greenbelt Bank Board, which was modeled after the South Carolina Conservation Bank, a recognized statewide authority that had a long record of achievement in land conservation. An urban grants program would be operated by the County Park and Recreation Commission and an Urban Grants Review Committee, which would be responsible for distributing funds to municipalities.

Local governments, private landowners, and land conservation organizations either in partnership or individually would submit grant applications to the County Bank Board or the PRC for funding to implement a conservation project. The county would staff the rural Greenbelt Bank Board, and the PRC would staff the urban grants program. The grants program was fully described within the approved Greenbelt Plan document in terms of what constituted an eligible project, the application and review process, and how funds would be awarded to priority-ranked projects. Criteria for ranking projects was also provided in the plan to guide the work of the Greenbelt Bank Board and the PRC staff.

Final Plan Approval

The final Greenbelt Plan report provided Charleston County with twenty-two specific action steps to complete, along with a specific time frame for completing those actions. Ten months had elapsed from the tour of the Francis Marion Forest in Dickie Schweers's pickup truck to the day the Charleston County Council accepted and approved the final Greenbelt Plan in June 2006. Much had been accomplished, and most important, the County Council had honored the will of the voters and established a greenspace conservation program that would in due time become one of the best in the nation.

Measuring Success

Success can be measured in many ways. For the Greenbelt program, the easiest way to define success is to look at the overall goals of the program and examine how the November 2004 sales tax funds were spent, how much land was protected and conserved, and how equitable the distribution of funds was across the county. From my perspective, Charleston County gets high marks on every measurable activity. The following table illustrates these measurements.

The other measure of success is found in the people who operate the program on a daily basis, the landowners who have participated in the program, and the citizens who support the mission and goals of the program. This begins with the work of Cathy Ruff, the Greenbelt program manager, who has dedicated the past twelve years of her career to faithfully executing the vision of the program. It takes a tremendous amount of hard work to keep up with a program that has amassed almost 40,000 acres of new conserved land in less than fifteen years. To many in Charleston County, Cathy Ruff is the face, heart, and soul of the Greenbelt program. Jennifer Miller also deserves credit for the success of the program. She now serves as county administrator and has been a proponent of the program since day one. The Greenbelt Advisory Board

GREENBELT COMPONENTS

Program	Acres Protected	Funding Awarded
Overall Greenbelt	21,170	$96 million
Rural Grants	17,374	$66.5 million
Urban Grants	772	$26.6 million
Small Landowner	12	$1.1 million

Status of Greenbelt Components (as of 2018)

	Goal	Total
Rural Lands (70%)	16,240	11,397
Francis Marion Forest (111%)	10,275	11,438
Lowcountry Wetlands (141%)	5,610	7,900
Charleston County Regional Parks (127%)	4,675	5,933
Urban Lands (33%)	2,000	666
Corridors (13%)	1,200	155
Total (94%)	40,000	37,489

has undergone almost 100 percent turnover and change, and there are many volunteers who deserve credit for the success of the program.

Authorization of Second Sales Tax, 2016

Perhaps the greatest measure of success occurred in November 2016, when Charleston County voters were once again asked to approve an additional sales tax that would generate an additional $2.1 billion in funding to support roadway development, mass transit, and greenspace conservation. Of the total funding, $210 million, or 10 percent of the sales tax, was for support of the Greenbelt program. Voters approved the referendum. County councilman Dickie Schweers was a proponent of the additional sales tax, stating that the Greenbelt program had a track record of success and that the additional funds would enable the county to continue the greenspace conservation work.

I believe that ten years of successful conservation swayed the voters to both trust the process and continue to support the Greenbelt program results. The Charleston County Greenbelt program is among the nation's most successful locally supported greenspace conservation programs. The program was established on the basis of a simple, yet compelling vision, which spoke volumes to Charleston County residents about the lands, water, and lifestyle that is critically important now and for generations to come.

Getting Ahead of Growth and Development

The lessons from the Charleston County Greenbelt program have tremendous value and application for numerous other rapidly growing communities and for coastal communities in the United States and abroad.

The citizens of Charleston County, without the benefit of a master plan, voted to tax themselves to improve their transportation system and conserve the Lowcountry landscape. Further, these citizens had faith that their local government would act in their best interests to develop a plan of action that would achieve the voter-approved vision. Charleston County is one of the fastest-growing metro areas in one of the fastest-growing mega regions in the United States. Citizens that voted, who are also the taxpayers, witnessed first-hand the changes in land use occurring in real time. It wasn't a matter of opposing growth and development; it was, however, a call to action to balance land development with conservation. The Greenbelt program has done nothing to harm the economy of Charleston County or diminish private property rights or values; in fact, the opposite is true.

One of the key takeaways from the Charleston County Greenbelt Plan is that American communities need to inventory the special landscapes that serve to define their sense of place. The places we live, the landscapes we call home, are critically important to the way in which we value our communities. The loss of special places is something many of us will experience in our lifetime. Sometimes the loss of place is from natural forces, such as hurricanes, tornadoes, fire, or flood. Other times the loss is due to human actions, such as land development or as a result of war. My friend Tony Hiss is the author of a book titled *The Experience of Place*, in which he defines the meaning of place in our lives as follows: "Our relationship with places we know and meet up with—where you are right now; and where you've been earlier today; and wherever you'll be in another few hours—is a close bond, intricate in nature, and not abstract, not remote at all. It's enveloping, almost a continuum with all we are and think." Hiss goes on to say, "These days people often tell me that some of their most unforgettable experiences of place are disturbingly painful and have to do with unanticipated loss."

Faced with the loss of their special landscapes, the citizens of Charleston County took action; they created a local funding program that supported a program aimed at saving the landscapes that they universally love. This is a model that other American communities should adopt and implement.

Climate Action Plans Needed

The term Global Climate Change has unfortunately been politicized and used to divide people into believers and nonbelievers. To be clear, we have never, at any time in human history, lived in a static global climate. The climate of our planet is dynamic and forever changing. Think for a moment to the climate of your youth and then consider the climate that you live in today. It does not take scientific study for everyone on planet Earth to understand that our climate is dynamic and changing.

I believe that those who identify themselves as deniers of global climate change are mostly concerned about the impact that acceptance of this as fact would have on the global and national economy and the way we conduct business. The concern is that we have to make great sacrifices to lower carbon emissions and reduce human impact to the air, land, and water. Embracing the fact that climate is dynamic could open new opportunities of economic growth and human advancement. We won't achieve these opportunities by living in denial of the facts.

One of the greatest future threats to American coastal communities is global climate change. Coastal communities like Charleston County must assess the impacts of rising sea levels and seasonal flooding on everyday life. A phenomenon known as "Sunny Day" flooding has been steadily increasing for the past five decades and impacting coastal communities such as Boston, New York, Philadelphia, Norfolk, Charleston, Savannah, and Miami. Programs like the Charleston County Greenbelt Plan are critically needed to protect public health, safety, and welfare by keeping residents and businesses out of landscapes that are prone to flooding and at the same time protecting the natural resource systems that are most capable of absorbing seasonal flooding. These are not political decisions; they represent common sense and appropriate action by vulnerable communities.

8

Callin' the Hogs

The Northwest Arkansas Razorback Regional Greenway, Arkansas

Cyclists ride along a portion of the Razorback Greenway north of downtown Springdale. Photo by author.

NORTHWEST ARKANSAS MAY NOT CURRENTLY be among your top ten list of must visit destinations. There is a chance you may have never even heard of Northwest Arkansas. You probably know or have heard of Walmart and Sam Walton, the American entrepreneur who founded the retail giant. Sam Walton was born in March 1918 in Kingfisher, Oklahoma, completed his college education at the University of Missouri in Columbia, and after World War II moved to Newport, Arkansas, where he purchased and operated his first Ben Franklin five-and-dime variety store. He and his wife, Helen, moved to Bentonville, Arkansas, in 1950 when Sam purchased a second Ben Franklin store located on the downtown square. From humble beginnings, Sam Walton transformed his collection of small-town variety stores into one of the world's largest corporations and America's largest private-sector employer.

After they amassed a considerable fortune, Sam and Helen established the Walton Family Foundation in 1987 with a mission to "tackle tough social and environmental problems with urgency and a long-term approach to create access to opportunity for people and communities."[1] The family foundation was "envisioned as a bond to strengthen and bring their family together, children and grandchildren across generations to find lasting solutions to big challenges." The foundation has made a number of strategic investments in Northwest Arkansas that have shaped community development and enhanced the lifestyle for residents and tourists, including the conservation of greenspace and development of new regional parks and greenways.

Sam Walton passed away in April 1992. His fortune and legacy were passed along to his sons and daughter, who not only assumed management of Walmart but also became stewards and leaders of the family foundation. One of Sam's sons, Jim Walton, along with his wife and sons Steuart, Tom, and James, began envisioning a future Northwest Arkansas that would be remarkably different from its rural past. They developed a passion for mountain biking and sought trips to Europe and popular American destinations that enabled them to ride a variety of single-track trails. It occurred to all of them that their home region possessed incredible natural beauty (Arkansas is called the "Natural State" for very good reason) and the rugged terrain necessary to build world-class mountain bike trails.

Tom Walton, who was chair of the family foundation's Home Region Committee (which invests in quality of life projects across Benton and Washington Counties, Arkansas) when I first met him in March 2010, began building mountain bike trail venues, with the first being Slaughter Pen situated just north of downtown Bentonville. Tom and his brothers worked to expand Slaughter Pen, and from this grew an interest in making the two-county region more bicycle friendly.

Today, Northwest Arkansas is among the fastest-growing metropolitan areas in the nation. This growth is driven in part by Walmart and two other Fortune 500 companies, Tysons and JB Hunt. When you visit Northwest Arkansas you can tour Crystal Bridges Museum to American Art, a world-class museum one block north of the Bentonville square, stroll through revitalized historic downtowns, eat at a variety of farm-to-table restaurants, ride mountain bikes on an extensive network of soft-surface single-track trails, and visit the original Ben Franklin five-and-dime on the square in downtown Bentonville, which has been turned into a museum dedicated to the history of Walmart. Most important, you can accomplish all of this on foot or by riding a bicycle, using the thirty-six-mile Razorback Greenway.

Vision for a Regional Trail

I began working with the Walton Family Foundation in spring 2010 when I was asked to submit a proposal for the preparation of a regional greenway master plan. I had barely finished reading the foundation's request for proposals when my phone rang. It was my good friends and colleagues Bob Searns and Jeff Olson, who were together at Bob's home in Colorado. "Did you receive a letter from the Walton Family Foundation?" they asked. "Yes," I responded. "Would you consider teaming with us to pursue this project? This is sure to be our next big adventure." I agreed, and together we formed "The Green Team" and submitted our qualifications to the foundation. Our mutual friend Andy Clarke, who at the time was executive director of the League of American Cyclists in Washington, DC, deserves credit for connecting all of us with the foundation. Andy had met with the foundation staff in 2009 to discuss their interest in bicycling and regional greenway development and provided our names as consultants capable of assisting the foundation.

The Green Team

Jeff Olson, who in 2010 was a principal and co-owner of Alta Planning + Design, headquartered in Portland, Oregon, reprised the role that he executed perfectly for the Grand Canyon Greenway—chief instigator, grand visionary, and team leader. The Green Team was composed of familiar faces: Jeff; Bob Searns; Bill Neumann, owner and landscape architect with DHM, a landscape architecture firm based in Denver, Colorado; and

Tom Woiwode, a friend, colleague, and former client with the Southeast Michigan Greenway Initiative, based in Detroit, Michigan. Our group partnered with local engineering firm, CEI, based in Bentonville, Arkansas. Our initial visit to Northwest Arkansas occurred on March 31, 2010, in Bentonville and included a kick-off meeting and two-day workshop with Walton Family Foundation staff, including home region director Rob Brothers, senior program officer Sandy Nickerson, human resources manager Janet Post, and Tom Walton. Andy Clarke also joined the initial Green Team meeting in Arkansas, serving to bridge relationships among the assembled group.

Building a Coalition

The purpose of our two-day workshop was to test the level of acceptance for a regional greenway and to develop a plan and strategy for building the project. Our team met with the Northwest Arkansas Regional Planning Commission (NWARPC), the regional metropolitan planning organization, led by executive director Jeff Hawkins and director of planning John McLarty, along with municipal mayors, business leaders from across the region, outdoor interest groups, and local government agencies. The team conducted an extensive walking, bicycle, and windshield tour of the two-county region to familiarize ourselves with the landscape and prepare an on-site assessment of the opportunities and constraints for regional greenway development.

In the process of completing our work, we came to better understand the decades of work that had already been accomplished by the Walton Family Foundation and NWARPC. The commission began working in 2000 to identify trails as a key component of the region's long-range transportation plan. At the same time, the foundation began funding municipal trail projects in Bentonville, Fayetteville, and Rogers. In that year, NWARPC identified three historic "trails" as the potential components for a regional trail: the Butterfield Stage Coach route, Civil War routes, and the Trail of Tears. These historic routes were combined with local trails to form what NWARPC called the Northwest Arkansas Heritage Trail. The Arkansas Highway and Transportation Department (AHTD) encouraged the communities of Northwest Arkansas to adopt and incorporate the planned route within their long-range transportation plans

Members of the design team, client, and municipal representatives gather around a conceptual map depicting the future route of the Razorback Greenway. Source: Author.

and projects. The downside of this prior planning effort was that much of the proposed regional trail would be constructed along heavily traveled roadway corridors, either as a signed route or as a bike lane. The reality of NWARPC's regional trail proposal fell short of the Walton Family Foundation's vision of an off-road regional trail that would be safe for a broader spectrum of the population to use.

Our team worked day and night for three days during that initial visit to stitch together an off-road route that stretched for thirty-six miles, extending through all six Northwest Arkansas municipalities: Bentonville, Rogers, Lowell, Springdale, Johnson, and Fayetteville. Our work was accomplished in collaboration with more than 100 participants. Upon completion, we presented our findings to an assembled group at a Bentonville hotel, where we received a polite if not slightly wary reception.

After all, it is easy to play with maps and colored markers; it is another task altogether to build an ambitious and complex multijurisdictional urban greenway.

After the intensive charrette work in Bentonville, the Green Team continued its work and during the months of April and May 2010 completed a series of four interrelated reports summarizing the results and findings of our work and the proposed "Northwest Arkansas Regional Trail." The reports defined a comprehensive vision and development program for the greenway, which we envisioned as an off-road, shared use trail that would extend from the Bella Vista Trail, in north Bentonville, south to the Frisco Trail in south Fayetteville. The trail, as planned, would link together dozens of popular destinations, including three downtowns, three regional hospitals, twenty-three public schools, the University of Arkansas, corporate headquarters of Walmart, J. B. Hunt Transport Services, Tyson Foods, arts and entertainment venues, shopping areas, historic sites, parks, playgrounds, and residential communities. The route included approximately 14.2 miles of existing trails and 21.8 miles of new trail that needed to be planned, designed, and constructed. The challenge of building a long-distance, off-road, shared-use greenway trail that linked together the communities of Northwest Arkansas depended on how the region would fund design and construction, as well as the long-term operations and management. Our team estimated that total project development costs would exceed $38 million.

Let's Call It the Razorback Greenway

As we were in the process of wrapping up the conceptual plan, Jeff Olson took exception to the working title "Northwest Arkansas Regional Trail." He believed that the name sounded bureaucratic and did not signify the importance of the greenway. Jeff suggested that the project be named the Razorback Greenway. Green Team members loved his idea. The only problem was the name was already trademarked by the University of Arkansas, where it had been used for decades. This was not an obstacle to Jeff. He had the perfect solution—simply contact the University and tell them about the significance of the greenway and the importance of using the name "Razorback" for the project. While the rest of us smiled and thought to ourselves, "Well, that will never happen," Jeff pursued his idea to the very top of the university administration. To our surprise, the university agreed with Jeff's suggestion, and the project was renamed "The Northwest Arkansas Razorback Regional Greenway." This called to mind one of Bob Searns's favorite sayings: "If at first you don't succeed, ask again *seven* times, until you get permission." For the purpose of naming the Razorback Greenway we only had to ask once.

Congratulations, You've Been Awarded a TIGER Grant

After we completed the final conceptual plan presentations, Rob Brothers approached the Green Team elated by the results and congratulating us on the success of our initial efforts. He asked: "Fellas, how would you recommend funding this project?" We looked at each other, and I responded, "Rob, there are lots of ways to fund these projects. Perhaps the best way would be to pool together local public funds and pair them with private sector funding." Skeptical of my response, Rob replied: "That sounds good, but we need something more realistic to work with. We need a funding proposal and we need it quickly to build on the momentum of the last few days." Then he added one final thought, "We don't want to wait twenty years to build this greenway. You're the experts, and you need to come up with a funding proposal that enables our region to have this project built within, say, five years." Rob put forward a serious challenge, and our team began thinking about the best way to respond.

One of our project reports focused on funding recommendations, and among those featured was a United States Department of Transportation (USDOT)–funded grant program called "TIGER," an acronym for Transportation Investment Generating Economic Recovery, which was part of President Barack Obama's "American Recovery and Reinvestment Act." TIGER was an important part of the president's 2008 stimulus program that was meant to get America's stagnant economy moving— creating jobs and providing much-needed domestic spending. TIGER was focused on making transportation improvements for projects that were "shovel ready" and could demonstrate an ability to stimulate a local economy.

We examined the possibility of applying for a TIGER grant, thought it was a long shot, and relayed our thoughts to Rob Brothers. The first round of TIGER funding (TIGER I) included $4 billion of stimulus

RAZORBACK GREENWAY

—— TRAIL   **TH** TRAILHEAD   521 MILE MARKER

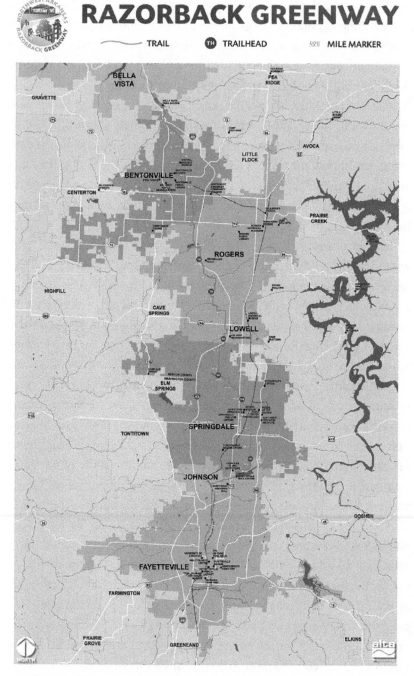

Final map of the constructed NW Arkansas Razorback Regional Greenway.
Source: Alta Planning + Design.

funding, in which two urban greenway projects were funded: $23 million for Philadelphia's urban greenway initiative, and $20.5 million for the City of Indianapolis Cultural Trail. We shared that information with Rob as well, and the foundation decided to apply for a TIGER grant, hiring Alta Planning + Design to author the application.

Jeff Olson took the lead within Alta to author the grant application. At one critically important luncheon at the Springdale Country Club, Jeff met with the region's movers and shakers and was able to secure forty letters of support for the TIGER application from those in attendance: local governments, businesses, congressional representatives, the governor's office, and state agencies. Jeff is a terrific salesman, and his work on the application paid huge dividends.

The Northwest Arkansas Razorback Regional Greenway application was one of more than 1,000 applications submitted for TIGER 2 funding from across the nation. The grant authored by the Green Team, the foundation, and NWARPC was one of forty-two capital projects awarded funding. Our grant application requested $24.8 million for design, acquisition of greenway right-of-way, and trail construction. In mid-October 2010, the Razorback Greenway was officially awarded $15 million. The Walton Family Foundation provided the required 20 percent "local" funding match of $3.75 million. (The municipalities in Northwest Arkansas did not contribute any funds to the TIGER grant as the foundation agreed to pay the required local match.) Total project funding was $18.75 million, which was approximately $6 million short of our initial estimate of construction costs. The federally funded portion of the Razorback Greenway was reduced from what was featured in the application to approximately sixteen miles in length, extending from the north end of Lake Fayetteville, in the City of Fayetteville, to the intersection of Interstate 49 and West New Hope Road in Rogers. The Northwest Arkansas Regional Planning Commission was the official TIGER applicant, as grant funds had to be awarded from the federal government to a local government agency. In addition to the work on the federally funded project, the Walton Family Foundation agreed to fund an additional $15 million of trail construction in Johnson, Fayetteville, and Bentonville, sections of proposed greenway not funded by the federal grant. In a matter of nine short months, the Razorback Greenway evolved from a vision and concept map to a fully funded project, thanks to the TIGER grant and the generous support of the Walton Family Foundation.

It should be noted that of the forty-two capital awards in TIGER 2, only two of the projects were greenways; the other forty were traditional highway, roadway, bridge, rail, water, and airport transportation improvements. Additionally, every application for TIGER funds had to address uniform federal selection criteria and define the long-term benefits of project development. The TIGER application that Alta prepared for the Razorback Greenway had to be consistent with long-term objectives defined by the United States Department of Transportation, and this included seven ways that "return on investment" (ROI) would be derived from project development:

Support livable communities that attract businesses and quality employees. This was a very important concern for the three Fortune 500 companies in Northwest Arkansas: Walmart, J. B. Hunt, and Tyson Foods. The Walton Family Foundation believed that building the greenway was key to maintaining the region's economic competitiveness. The following quote from the foundation was featured in the TIGER application: "The Razorback Greenway will be good for the Northwest Arkansas economy and the health of our people. Healthy people and a good economy mean good business."

Attract and enhance dining and lodging businesses in the region. The goal of the greenway was to attract more visitors to the region and extend time visitors spent in the region. Tourism in Northwest Arkansas was the second-most important economic factor in the region, generating $690 million in annual revenues. Benton and Washington Counties' tourism revenues were ranked among the top five in Arkansas. Enhancing tourism was an important economic outcome of the greenway.

Increase real estate values and sales. Studies conducted by the National Homebuilders Association and National Association of Realtors have consistently shown that living within close proximity to parks, greenspace, and trails improves property values—in many cases by as much as 20 percent above comparable properties. Razorback Greenway sponsors hoped to realize that return on investment for property owners across the region.

Increase outdoor recreation equipment and apparel sales. This was another economic benefit that was of interest to the foundation and other project sponsors. It was one of the first economic benefits to materialize as new bicycle retail shops opened in both Bentonville and Fayetteville before the project was completed.

Improve health and fitness lifestyles, helping to reduce medical treatment and costs. With so much focus, debate, and angst regarding the federal Affordable Care Act, the Razorback Greenway application offered a community trail that could be used daily to improve the fitness and health of Northwest Arkansas residents. The simple prescription: walk, bike, or run for a healthier life. Mercy Hospital in Rogers immediately bought into the health benefits of the greenway and agreed to have a trailhead constructed on a portion of the hospital property.

Increase fiscal revenues through a healthier tax base. Our TIGER application demonstrated that the greenway would generate more revenue from a variety of sources and as a result would increase tax revenues. Generating a stronger tax base would enable municipalities of the region to have more funding to address a variety of community needs.

Create jobs in the areas of construction, trail operations and management, and the service industry across the region. We forecasted that the greenway would create approximately 390 new jobs within five years of completion. I conducted an informal survey in 2015 and found that approximately 200 of those jobs, directly tied to greenway development, had already been created prior to completion of the regional trail.

The Legacy of Transportation Funding

It may seem odd that a trail project is funded with federal transportation dollars. During the past four decades federal transportation funding has been critically important to building thousands of greenways, rails-to-trails conversions, and bicycle and pedestrian projects across the nation. There are detractors who believe that funding trail development with federal transportation dollars is a misappropriation of that money. I strongly disagree with that conclusion and have witnessed how the funding of walking and biking projects has significantly improved transportation efficiency, mobility, and public safety within rural and urban communities across America.

The decision to use federal transportation dollars in support of bicycle and pedestrian facility construction changed dramatically with passage of the Intermodal Surface Transportation Act of 1991 (ISTEA). Signed into law by President George Herbert Walker Bush on December 18, 1991, ISTEA was the first major transportation policy initiative of the post Interstate Highway era to dedicate funding to bicycle and pedestrian projects. Prior to ISTEA approximately $7 million was being spent nationwide on bicycle and pedestrian projects. During the six-year life of ISTEA $1.4 billion was spent, a rate of almost $400 million per year.[2]

ISTEA was a major change in philosophy for surface transportation funding, promoting all modes of travel, with specific focus and consideration for walking, bicycling, and transit projects. ISTEA provided critical financial support for converting abandoned railroad corridors into trails, funding the first such trail in 1995, the Cedar Lake Regional Trail in Minneapolis. ISTEA also elevated the role of metropolitan planning organizations (MPO) in the development of multimodal regional transportation plans. The Northwest Arkansas Regional Planning Commission was the MPO for the two-county region and took the lead role in overseeing development of the Razorback Greenway. This was possible due to the passage of ISTEA.

The Brookings Institution, a nonprofit public policy organization based in Washington, DC, that conducts in-depth research on a variety of subjects, accurately described the important role that ISTEA had on the development of surface transportation in the United States, concluding that "ISTEA is rightly considered to be a watershed moment in the U.S. transportation policy. It offered a new framework for thinking about transportation by assuring states and metropolitan areas of specific levels of funding and giving them the flexibility needed to design transportation mixes that met their needs."[3]

ISTEA was the beginning of new attitudes and approaches to the use of federal transportation funds, and when it expired in 1997 Congress replaced it in 1998 with the Transportation Equity Act for the 21st Century

(TEA-21). When TEA-21 ran its course, it was replaced in 2005 with passage of the Safe, Accountable, Flexible, Efficient Transportation Equity Act: A Legacy for Users (SAFETEA-LU). The most recent major update to the surface transportation policy occurred in 2012 with passage of Moving Ahead for Progress in the 21st Century Act (MAP-21). The result of these successive federal transportation policies and programs has been significant financial support for the development of trails, greenways, and bicycle and pedestrian facilities. According to my calculations, the United States has added approximately 50,000 miles of new trails across America since the passage of ISTEA, which is equivalent in size to the completed Interstate Highway System.[4]

Saving the TIGER funds

Shortly after the TIGER grant award was made in November 2010, the Green Team, Walton Family Foundation, and NWARPC were engaged in a flurry of activity to save the funding. In January 2011, the U.S. Congress undertook an evaluation of federal funding and put forward a proposal to shut down the TIGER program and rescind unobligated funding. From January through the end of March 2011 the NWARPC reached out to the Arkansas governor's office, as well as Arkansas congressional delegation and business leaders to reinforce that the region wanted the TIGER grant, needed the money to complete the Razorback Greenway, and was counting on Washington, DC, to fulfill its commitment to fund project development. The Walton Family Foundation organized a busload of Northwest Arkansas residents to attend congressional hearings that were being conducted in Tulsa, Oklahoma, to oppose rescission of the TIGER funds. I was asked by the foundation to attend two meetings that took place in Little Rock with the Arkansas Highway and Transportation Department and Federal Highway Administration officials to advocate for the funding.

At an April 6, 2011, hearing on National, State and Local Transportation Priorities for the Next Surface Transportation Authorization, Mike Malone, president and CEO of the Northwest Arkansas Council (a private nonprofit founded in the 1980s by business and civic leaders Sam and Alice Walton, Don and John Tyson, and J. B. Hunt, to identify regional challenges and serve as a catalyst for economic development solutions), testified before Senators Barbara Boxer and James Inhofe and other members of the United States Senate Committee on Environment and Public Works on behalf of full funding for the Razorback Greenway. Malone's statement included the following: "This [Razorback Greenway] $30 million-plus project received a $15 million TIGER II grant in October of 2010. The Federal investment will leverage another $15 million of private funding from within the region. It is the perfect example of how public agencies and private sector interests can work together to make a difference. However, there is presently great concern over the House of Representatives' budget proposal that would rescind all unobligated TIGER II funds—including the $15 million committed to the Razorback Greenway. A large amount of planning and preliminary work has occurred already based on the TIGER II award announcement. Withdrawing previously pledged funding could put the private sector funds at risk and would certainly delay this project for many, many years."

All the hard work paid off. Late in the afternoon of April 8, 2011, NWARPC received official notice that the TIGER funding had been "obligated," a term that in effect meant the funds were deposited into a federal account and therefore could be spent on the project without risk of rescission. It was an exhausting four months of effort in which every day felt like the most important day to save the TIGER funds. Thankfully, that phase of our work was over. It was time to move ahead with design and construction of the greenway.

Greenway Design Begins

In January 2011, the NW Arkansas Regional Planning Commission decided that in order to meet the timeline required for spending the TIGER funds, it had to move forward with selecting a consultant to prepare the construction drawings, specifications, and cost estimates for the federally funded portion (sixteen miles) of the Razorback Greenway. Once the TIGER funds were obligated in April, and after an exhaustive review of almost a dozen professional submittals from a plethora of national firms, the NWARPC recommended hiring the Green Team, led by Alta Planning + Design.

The official notice of selection occurred on May 15, 2011. Finalizing the design contract took another two months. At sixteen miles in length and

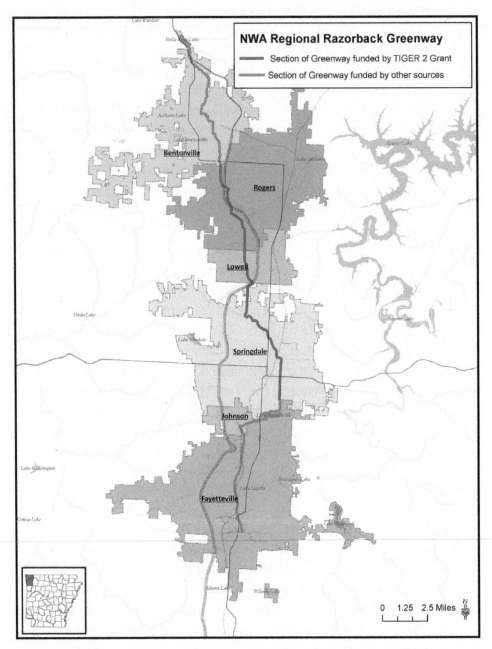

Map depicting the federally funded portions of the Razorback Regional Greenway. Source: Northwest Arkansas Regional Planning Commission.

with a price tag of $38 million, the project was too large to design and construct in a single phase. The Green Team first generated a more detailed route and alignment plan for the entire greenway, walking the entire route again with the foundation staff, NWARPC, and municipal leaders. After the site walk, we produced four illustrated conceptual master plan maps. We further subdivided the greenway into fundable segments, in the range of $3 to $5 million in value, to successfully advertise bids and award construction contracts. This was a bit tricky as each segment had to have an independent transportation function, which meant there had to be logical end points that could serve a travel and transportation purpose, and we had to connect to other existing transportation systems, such as roads, sidewalks, and bicycle facilities. There was legitimate concern that, even with the TIGER grant in hand, we might run out of funding and therefore not be able to build the entire sixteen-mile federally funded project. As a result, the project was subdivided into nine separate phases of construction.

When the Green Team's original design proposal was submitted to NWARPC and the Walton Family Foundation, Alta Planning + Design and CEI Engineering were tasked with completing all the design and engineering work for the entire sixteen-mile TIGER-funded project. NWARPC and the municipal governments required that local engineering firms participate in the design of the project. The mayors of each municipality had ideas on how the work should be doled out and wanted to see local businesses working on the project. The team grew to include the following companies: Crafton and Tull, Rogers; USI Engineering, Springdale; McClelland Engineers, Fayetteville; Cook and Associates, a real estate firm in Fayetteville; Rife and Associates, a real estate appraiser based in Fayetteville; and Reed and Associates, a real estate appraisal firm based in Rogers. The other design partners included original Green Team members Bill Neumann and Bob Searns. It became an enormous team with lots of moving parts, which added to the complexity of the project. One of my jobs was keeping track of everyone's work program, subcontracts, payments, and deliverables.

Securing Land for the Greenway

At the outset of our work program, we sought to incorporate as much public input as possible. We wanted to hear from the citizens of Northwest Arkansas and see if they agreed with where the Green Team thought the greenway should be constructed. In August 2011, we conducted a series of public open-house workshops in Springdale, Lowell, and Rogers. We advertised the meetings in English and Spanish and did our best to encourage citizens from minority communities to attend the meetings. The turnout at our meetings was less than we hoped; however, we connected with several landowners who attended the meetings to see how the greenway affected their property.

One of the thorny issues that we had to address immediately during the design phase was the acquisition of greenway right-of-way. For the off-road portion of the project, we needed a minimum thirty-foot right-of-way to build and operate the greenway. For sections of the trail that ran parallel to roadways, the existing road rights-of-way were wide enough to accommodate trail development, but in some cases we needed to acquire additional land, and at times fifteen feet of right-of-way from the edge of the existing roadway corridor, to build the trail. The need for this right-of-way along roadways, as well as our need for new rights-of-way for the off-road trail, meant we would need to negotiate land acquisition with numerous property owners. As the Razorback Greenway was federally funded, we had to complete our acquisition work according to the federal government's Uniform Relocation Act. This process is used for all federally funded transportation projects and defines a formal, lengthy process that requires every property owner be approached and treated in the same manner. For the Razorback Greenway, there were 122 separate properties located in three different municipalities that were included in the right-of-way acquisition process. Virtually all those properties represented a unique acquisition strategy, meaning we could not simply duplicate successful negotiations from one property owner for use with another. Each property provided a unique set of circumstances that had to be resolved and documented. We had to provide evidence of all communications and documentation in triplicate, with one set of files for NWARPC, one set for the municipality, and one record for our files.

Dennis Blind, ASLA, of Alta Planning + Design, who in addition to being a registered landscape architect was also an Arkansas-licensed real estate agent, and I worked together to complete the negotiations and acquisitions for all 122 properties. It took more than a year to accomplish, and it was essential work for the project to be successful. Twelve property owners disputed the thoroughness and accuracy of our assessments and subsequent purchase offers and exercised their right to challenge our work in a court of law. Working through the legal process meant additional time and money; however, in the end, all twelve property owners agreed to out-of-court settlements that enabled us to secure the necessary right-of-way to build the greenway.

Daylighting Spring Creek

The longest federally funded section of the Razorback Greenway is located in Springdale, which at one time was the largest and most important municipality in Northwest Arkansas. Years before Sam Walton put Bentonville on the map with the founding of Walmart, and before the University of Arkansas transformed Fayetteville into the region's largest community, Springdale was the hub of life, commerce, and activity in Northwest Arkansas.

From the 1930s through the 1960s, Springdale was the agricultural marketplace community, where area farmers would bring harvested fruits and vegetables for wholesale and retail sales and distribution. Springdale is also the birthplace of Tyson Foods Inc., a multinational corporation founded in 1935. During a particularly rough year for agriculture and crop farming, John Tyson began his chicken hatchery in downtown Springdale, which he eventually grew into a successful business enterprise. At the time we began work on the Razorback Greenway, Springdale was a shadow of its former existence. The downtown consisted of abandoned and dilapidated buildings. Downtown Springdale would have been a ghost town were it not for the municipal offices located one block from the intersection of West Emma and North Main Streets.

During the Green Team's initial conceptual master plan work in April 2010, Jeff Olson asked that I join him for a walking tour of Springdale to help define the future route and alignment for the Razorback Greenway.

Before we began the walk, we were cautioned by local members of the Green Team that building a greenway through Springdale would be the most challenging undertaking of the entire project. Despite being one of the most populated communities in Northwest Arkansas, Springdale had few parks and less than a mile of paved trail. To make matters more complicated, the route of the greenway would have to make use of what we refer to in the greenway business as "overland" routes—greenway trails that don't follow a natural drainage area, creek, stream, or river, but instead cross over the top of ridgelines and therefore must make use of existing rights-of-way such as road, railroad, or utility corridors.

It was raining and very cold as we walked from the south end of Springdale toward the downtown. We were navigating along roadways trying to find the most suitable greenway route and determining if we had enough right-of-way on either side of the road to build a trail. As we approached the center of the city, I noticed that our travel route intersected an urban stream, named Spring Creek. It was an ugly-looking stream that had been cleared of vegetation and shaped into an open channel; it was more of a ditch than what one might consider to be a stream, but it had flowing water, and we followed it into the heart of the downtown. Before we reached the downtown, the open creek channel disappeared into a gaping concrete culvert that carried the stream through the center of town.

We decided to take a break from the weather and took shelter in the only downtown restaurant open at that time, the Spring Street Grill. As we were drying out and warming up, it occurred to me that downtown Springdale was the only community of the six in Northwest Arkansas with flowing water, despite the fact that the creek had been covered over with concrete and was not visible throughout the downtown. I made a bold statement to my team that one day, downtown Springdale would be the "jewel of the Razorback Greenway." Asked why I thought this would happen, I cited the presence of the creek. Several team members quickly reminded me that it was encased in a concrete drainage culvert, to which I replied we would simply have to remove the concrete and daylight the creek. The team looked at me as if I were crazy and immediately dismissed my comments. We had so much challenging work ahead of us just to build the Razorback Greenway that the thought of transforming Springdale's downtown was low on the list of priorities.

To be honest, I understood why there was such an underwhelming response to my comments. Downtown Springdale was dying. I could tell that some effort had been made to renovate and revitalize the downtown, but by appearances that too had failed. It would take a lot of effort to revitalize and revive the center of the city. As we began work on the Razorback Greenway, I never forgot that initial walk or my first impressions of downtown Springdale.

On August 1, 2011, we were conducting public outreach meetings in Springdale, and I realized that there was an opportunity to revisit my vision for downtown Springdale revitalization. I had lunch earlier in the day with Patsy Christie, the city's hard-working and visionary planning director, who had worked with us on the greenway since we first arrived in March 2010, and brought up the idea that the Razorback Greenway could be a catalyst for the revitalization of the downtown. Patsy was intrigued and asked how I envisioned this occurring. I shared with her my proposal, which involved demolishing the concrete box culvert that was covering Spring Creek through the middle of the downtown. Through some research, I learned that the culvert had been constructed in the 1970s as part of a flood control and urban renewal project. My plan included daylighting the creek to create a new landscape, with Spring Creek becoming the focus; revitalization of the center city would follow. Patsy enthusiastically supported my idea and encouraged me to share my vision with Springdale mayor Doug Sprouse. Later that evening, during the public outreach meeting, I pulled Mayor Sprouse aside and hastily drew a sketch of my proposal. The mayor immediately sat down, looked at me, and said: "I have never given much thought to uncapping the creek and making it the centerpiece of downtown revitalization. This is an interesting idea. Can you prepare a more fully developed proposal?"

A couple of weeks later, the Green Team assembled in Springdale to generate a conceptual master plan for the revitalization of the downtown that involved daylighting Spring Creek. I conducted additional research into the history and significance of Spring Creek and found that it was the namesake for the community (just like the Miami River in Florida). The original name for the community of Springdale was a biblical reference—Shiloh. In the late 1800s to early 1900s Spring Creek was used for public baptism, and the community that formed around the banks of the creek was given the name Shiloh. Later, the community was renamed Springdale in honor of its natural heritage. Somehow, over time, this heritage had been forgotten by many who lived in the community.

Patsy Christie worked to form a new group to champion the concept of downtown revitalization. In the fall of 2011, I met with the newly formed Downtown Springdale Alliance. Serving on the committee was Walter Turnbow, born in 1924, who was elder statesman and a determined voice for downtown revitalization. Turnbow was not happy that city residents had allowed the center city to deteriorate. He recalled playing in Spring Creek as a child and loved the idea of daylighting the creek as the centerpiece of downtown revitalization. Walter was a vocal member of the alliance and would often chide fellow members during their meetings to "do something before I die."

Through the fall and winter months, I made several presentations to the alliance, sharing with them more specific design concepts and ideas for making revitalization a reality. The alliance wanted to complete a master plan for uncapping Spring Creek but wanted it framed in the context of a plan for the entire downtown. The work on the Springdale Downtown Master Plan took a few months to complete and produced a viable development strategy for the community. Most important for me, the plan provided a framework for daylighting Spring Creek and jump-started the transformation of the central city.

Daylighting Spring Creek proved to be more challenging than anyone could have imagined. It included removing two very old buildings, closing an existing downtown street, and creating new civic space where none existed before. The Walter Turnbow Park is the result—a center city park that provides a new identity for the community as a thriving environmental, commercial, and events space that serves to promote renewed life in this historic community.

In 2017, this is how the Downtown Springdale Alliance now described its vibrant city center: "Located along the Razorback Greenway and lovely Spring Creek, Downtown Springdale is packed with stylish boutiques, dynamic cultural institutions, fantastic eateries, and a burgeoning brew scene, all in the heart of gorgeous Northwest Arkansas." Trust me when I say that nothing even close to this would have been said of downtown Springdale in spring 2010. In seven short years, downtown Springdale has been transformed from decaying and dying to a vibrant city center.

Photo depicting the site of Walter Turnbow Park prior to construction. Photo by author.

Photo depicting Walter Turnbow Park after removal of box culvert and construction of park. Photo by author.

Revitalizing downtown Springdale was not the only challenge we faced in finding a suitable route for the Razorback Greenway through the community. An equally daunting challenge was finding enough public right-of-way along the "overland routes" in the southern and northern portions of the community capable of supporting trail construction.

As Jeff Olson and I completed our initial walk in April 2010, we met with Green Team members Bill Neumann and Bob Searns to discuss possible design solutions for two lengthy sections of the Springdale portion of the Razorback Greenway: South Powell Street and North Silent Grove Road corridors. These routes were selected as the preferred corridors for greenway construction, and there were not a lot of alternative routes to consider. The problem with both routes was the constrained rights-of-way that made constructing a high quality shared-use greenway trail very challenging. Our team proposed building what was at the time a new urban greenway design strategy called a cycle track (which involved converting a portion of the roadway for use as a bicycle path, merging the experience of a trail with the in-street experience of a bike lane). Few completed cycle tracks existed in the United States at the time that we proposed installing them as part of the Razorback Greenway. To make matters more challenging, nothing resembling a cycle track had ever been built in Arkansas.

I was tasked with crafting a white paper, which was a technical design brief that advocated for the construction of two, two-way cycle tracks in Springdale. With substantial help from the engineers at Alta Planning + Design, the paper was submitted to the Arkansas Highway and Transportation Department and the USDOT Federal Highway Administration regional office in Little Rock. Initially the response to the idea and white paper was "Nope, no way." Cycle tracks were a concept and were not codified under federal transportation policy. The Obama Administration was supportive of the idea of cycle tracks, but there was insufficient performance data about constructed cycle tracks to warrant their use. This was of particular concern for TIGER-funded projects as these were not landscapes intended to feature experimental design. TIGER emphasized "shovel-ready" projects, and cycle tracks were not considered a shovel-ready solution. The initial response from AHTD and FHWA was that

Schoolchildren ride their bicycles on a two-way cycle track on Powell Street in Springdale, Arkansas. Source: Nancy Pierce photographer/Alta Planning + Design rights holder.

cycle tracks needed more vetting and proven results before they could be included in TIGER-funded project development.

The City of Springdale joined with NWARPC to advocate for the construction of cycle tracks and agreed to the maintenance and operation of them as part of their transportation network. After a series of detailed discussions involving design and construction, AHTD and the FHWA regional office reluctantly approved the use of cycle tracks for the Springdale route of the greenway. We were successful in advocating for, designing, and building an innovative solution for the Razorback Greenway. Those two sections of cycle track in Springdale have been vital to the success of the greenway. Without them, we would not have been able to complete the project.

The Art of Greenway Design

Where we did have unconstrained landscapes to work within, we employed more creative design solutions that resulted in the development of the most memorable and popular sections of the Razorback Greenway. Variety, rhythm, and syncopation are three words I always keep in mind as I lay out and design greenway trails. The way in which a trail is

An iconic section of the Razorback Greenway, south of Lake Springdale. Source: Nancy Pierce photographer/Alta Planning + Design rights holder.

built on the land—how it responds to the shape of the land, surrounding vegetation, and other natural features—influences the trail user experience. Good trail design is intentional, and there is one stretch of the Razorback Greenway that has become one of the most enjoyable for public use due to the art and aesthetics of design and construction. This section is located north of downtown Springdale, and south of Lake Springdale, where we routed the Razorback Greenway through a rural agricultural field, adjacent to Spring Creek. There were two important challenges in designing this stretch of the Greenway: (a) acquisition of right-of-way and (b) trail layout and design.

After the team determined a rough alignment for the trail, we worked with two property owners along this stretch of the greenway to resolve conflicts between trail location and private interests. The largest tract of land was owned by a gentleman farmer who was leasing it for cattle grazing. His concern was that the greenway would compromise the safety of his livestock grazing operations. We assured him that greenway development would not interfere with his cattle and shared with him our desire to locate the greenway along the edge of Spring Creek so we did not encroach on his pastureland. We proposed separating the trail with a three-rail board fence facing the trail and an adjacent barbed wire fence facing the pasture. The dual fence solution was expensive, but it addressed important safety issues for each activity.

A second property owner presented different challenges, as their home was perched on a hill overlooking Spring Creek. Originally, our alignment of the greenway was parallel to Spring Creek and with an unobstructed view of their home. When I visited with the property owners at their home, they invited me onto the rear deck, which overlooked Spring Creek. The greenway would have been constructed in a manner that in my opinion impacted their viewshed and privacy. I recommended we move the greenway away from the creek and reroute the trail along the far edge of their property where it was out of view. The property owners were skeptical at first but agreed that our revised trail route was better than the original alignment.

With respect to the aesthetics of the trail, we wanted the alignment to have a loose free-flowing design that would fit with the natural contours of the landform and mimic the natural meander of Spring Creek—again utilizing variety, rhythm, and syncopation. We worked with the local engineering firm, USI Engineering based in Springdale, to complete an alignment that complemented the land and respected the rights of the property owners.

One of the most important features of this stretch was a bridge crossing over Spring Creek. During our initial field walk of the route, we discovered a section of raised limestone in Spring Creek that during high water flows created a small waterfall. Bill Neumann prepared a concept sketch that illustrated the design intent for this area. We envisioned building a bridge adjacent to the limestone cap so that trail users could view the exposed natural rock feature as they crossed the creek.

We wanted to be certain that we aligned the greenway trail as close as possible to Spring Creek because we wanted trail users to be close to the natural water feature, and we wanted the greenway to form a hard edge against the pasture so that trail users would have a view across the expanse of the open agricultural field. Our design strategy worked and is one of the reasons that this stretch is one of the most beloved along the entire thirty-six-mile length of the Razorback Greenway. This stretch pays homage to the rural heritage of Northwest Arkansas and is one of the most peaceful stretches of the trail.

Building the Greenway

More than two years elapsed before actual greenway construction commenced, and once it was under way, it took approximately three years to finish construction. To guide the construction process, first the Green Team prepared a set of design development standards for the entire sixteen-mile federally funded project. This involved deciding the most common and essential construction elements that would occur for the greenway project, such as trail tread construction (the preference was to build a concrete trail), bridges, drainage structures, regulatory signage, and landscape planting details. With the details package complete, each engineering team was provided with a set of guidelines and asked to generate construction document submissions at the 30 percent, 60 percent, and 90 percent completion stages.

As mentioned earlier, the project was divided into nine separate phases. The phases were not bid sequentially, but rather the natural order of completed design, permitted and approved construction documents, and shovel-ready status dictated the order of construction. As a result, the first two phases let for bid were 7 and 9, both located in Rogers. Phase 9 was already famous among Northwest Arkansas residents, politicians, and greenway advocates as it was highly visible from Interstate 49, which bordered one side of the project corridor. Phase 9 featured a partially constructed bicycle and pedestrian bridge, sitting by itself, up in the air on concrete abutments spanning Blossom Way Creek. The bridge was affectionately known throughout Northwest Arkansas as "the bridge to nowhere." At the formal greenway ground-breaking ceremony in the fall of 2011, I declared to trail supporters gathered that day that the "bridge to nowhere was now the bridge to the future."

Phase 7 was a complicated section from the outset. Due to the threat to TIGER funding, the foundation and NWARPC decided it was critical to push Phase 7 into construction. I had voiced concerns that the Phase 7 design was substandard. Nevertheless, the decision was made to move ahead. I worked with Crafton and Tull, designers and engineers for the project, to modify the design somewhat to improve its function. Phase 7 spanned beneath Interstate 49 bridges, and it had to make use of a narrow sidewalk along New Hope Road, the far right lane of which served as a connection to the Interstate on-ramp. It was not a good pedestrian and bicycle connection, and everyone knew it. Two years after it was constructed, the foundation hired Alta Planning + Design to redesign the connections in this area. I proposed a very expensive and highly visible Y-shaped bridge as an off-road solution to this challenging landscape. Eventually, this solution was built and opened to public use in 2017.

Following the completion of Phases 7 and 9, other phases of construction were undertaken concurrently. At one point in the process, we had seven phases of construction under way simultaneously. The order in which the phases were let for bids made no sense to the outside world, but it made perfect sense to me. Here is the order in which the remaining phases of the project were awarded:

Phase 4A—April 2012
Phase 4B—October 2012
Phase 3B—May 2013
Phase 5—July 2013
Phase 3A—September 2013
Phase 1—November 2013
Phase 2—February 2014

All of this kept me, the client team, and the Green Team extremely busy. I had to keep up with the status of construction work in three different jurisdictions (Rogers, Lowell, and Springdale), make certain that construction was adhering to the construction documents, process change orders for all seven phases of construction, and assist John McLarty and the NWARPC team in processing millions of dollars in payments to contractors. Most important, I had to keep track of all the project money, to the penny, and assure the foundation that we had enough funds to finish construction.

Tears of Joy

At the beginning of January 2015, we entered the home stretch of building the Razorback Greenway, and I knew that all phases of the project would be completed soon. John McLarty and I began to prepare official handoff of the greenway from NWARPC to each of the respective municipal governments for operation, maintenance, and management. The transfer of finished greenway segments from NWARPC (as builder) to

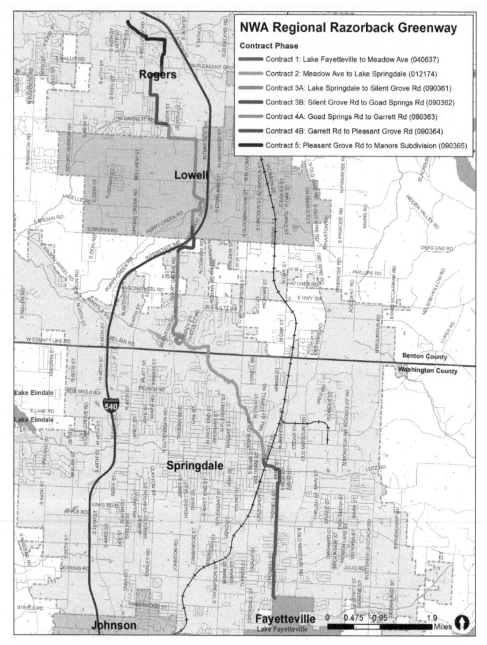

The Razorback Greenway phasing map illustrates the manner in which sections of the green-way were constructed over a three-year period. Source: Alta Planning + Design.

the municipal governments (as owners and operators of the greenway) activated a Governance, Operations and Management Plan that each municipality formally adopted as public policy.

At the same time, NWARPC, the foundation, and local governments began planning for a grand opening of the greenway that would be held at the beginning of May, settling on Walter Turnbow Park in downtown Springdale as the event location. This was poetic justice as far as I was concerned. In the spring of 2010, the City of Springdale was widely regarded as the community where we would have little success with greenway construction. Five years later, Springdale was the host community for the official grand opening of the Razorback Greenway. My prediction that Springdale would become the "jewel of the Greenway" had come true.

Those four months in 2015, from January to May, were incredibly stressful. There was lots of construction work yet to complete and, simultaneously, important preparations required to hold an official grand opening. It was just as the old saying goes that "the last 5 percent of every construction project requires 95 percent of your effort." We needed one final major push to meet our deadline, so I reached out to the construction contractors and engineering firms that were responsible for building each phase, encouraging them to finish their work as soon as possible.

The morning of May 2, 2015, arrived. It was a glorious day, with clear blue skies, abundant sunshine, and mild temperatures—an incredible day for a bike ride and a grand opening celebration. The Green Team assembled during breakfast at the 21c Museum Hotel, then made our way across Bentonville Square to Phat Tire to collect our rental bikes. An entire day of activity had been mapped out by NWARPC and the municipalities. I could tell right away that every member of the team was emotional as we began riding along the completed greenway, from downtown Bentonville south toward Springdale. We were giddy, proud, and in awe of the constructed greenway. It was approximately a seventeen-mile bike ride from the hotel to Walter Turnbow Park in Springdale. We enjoyed every foot of that ride.

I arranged for the Green Team to meet Rob Brothers on a section of the Razorback Greenway that is located immediately across the street from his home in the community of Lowell. When we met Rob, I could tell by his facial expressions that he too was filled with emotion. Every member of our team was proud of what we had accomplished. Rob, as the principal representative of the Walton Family Foundation, shouldered the expectations of the job as much as anyone. He had recently recovered from successful hip replacement surgery and had taken seriously the fact that the greenway offered him vastly improved, off-road, nonmotorized travel to points north and south of his home. The greenway had become an important part of his daily health and wellness routine. Rob was a living testament to the benefits of the greenway. He looked at the Green Team members assembled along the greenway and said: "Thank you for all that you have done to make this project a reality. Let me add, with all sincerity, that you changed my life—for the better." His voice cracked as he spoke those words, and we knew that his sentiments were from the heart.

As we continued our ride into downtown Springdale, I noticed that the traffic on the trail began to increase. The ride was building to a crescendo. Our team entered downtown Springdale north of Turnbow Park, and the level of excitement grew exponentially. As we pulled into the park, we could not believe the size of the crowd that was assembled. Having been part of other grand opening celebrations, I thought this was one of the largest that I had personally witnessed. More than one thousand enthusiastic souls were mingling throughout the park. A massive bike-parking station was set up in the middle of Emma Street, one of the major east-west routes in downtown Springdale, which had been closed to automobile traffic to accommodate the celebration crowd. There were so many people at the dedication ceremony that it was hard to navigate through the crowd to park our bikes.

Andy Clarke and I were invited to address the assembled crowd. Right before we were set to take our place behind the podium, a woman came rushing out of the crowd, threw her arms around my neck, and said to me: "Chuck, thank you for what you have done here in Northwest Arkansas." It took me a second to realize who was speaking to me. It was Marilyn Morton, one of the front desk clerks with the Doubletree Hotel in Springdale. The Doubletree had become my home away from home during my six years of work in Northwest Arkansas. Marilyn and hotel manager Steve Wright were enthusiastic supporters of the greenway project and always glad to see me when I arrived in town for work. I thanked Marilyn for her kind words and years of support. "Chuck, look at me, I have lost weight. I had not ridden a bicycle in years. Now I ride my bike two to three days each week. You have changed my life—for the better. Thank you so much!" She had tears streaming down her face. That was the second time in approximately one hour that a resident of Northwest Arkansas talked about their lives having changed for the better.

The ceremony began with the mayors from all six communities addressing the crowd. John McLarty spoke on behalf of NWARPC, Karen Minkel, the new home region director (Rob Brothers had retired), spoke on behalf of the Walton Family Foundation, and so on. It was a long ceremony. As for my part, I talked about the unsung heroes of the greenway—the property owners who graciously worked with us to provide the right-of-way needed to build the trail. Without landowner cooperation the project would never have happened. The last speaker of the day was Arkansas Highway Commissioner Dick Trammel, who led the enthusiastic crowd in a stirring rendition of "Calling the Hogs," a University of Arkansas cheer that is used to rally the Razorback sports teams but, as I now discovered, was also used to celebrate other achievements. The cheer goes something like this: Wooooooooooo, Pig! Sooie! Wooooooooooo, Pig! Sooie! Wooooooooooo, Pig! Sooie!—Razorbacks!

With the formal part of the ceremony complete, the crowd gathered along the edge of the completed trail in Turnbow Park and cheered loudly as a group of students rode their bicycles from north to south, breaking a yellow ribbon that had been stretched across the trail. It was official; the Northwest Arkansas Razorback Regional Greenway was open for public use.

It wasn't long after that brief celebratory ride that the crowd began to disperse. The Green Team gathered for one final photo on the Springdale boardwalk north of Turnbow Park. It offered me time to reflect on six years of work. My thoughts immediately returned to the Walton Family Foundation's original stated purpose for the project: "The Razorback Greenway will be good for the Northwest Arkansas economy and the health of our people. Healthy people and a good economy mean good business."

Standing on the northern edge of Turnbow Park, it occurred to me that those words rang true. This project was a fantastic addition to the communities of Northwest Arkansas. Through personal testimony and observation, it was clear that the greenway was improving the lives of community residents, who were living more active lives, and this new lifestyle was having a measurable impact on their health. The greenway was already improving the economy of Northwest Arkansas. One only needed to look at a revitalized downtown Springdale to appreciate the economic benefit. Three of the mayors described an increase in tourism that they were witnessing in their communities as a result of the completed greenway. The Razorback Greenway was helping to put Northwest Arkansas on the map. The project was contributing to placemaking and created a signature amenity for the regional community. It was clear to me that healthy people and a healthy economy were the primary outcomes of the Razorback Greenway. The objectives had been fulfilled and our mission accomplished.

Lessons Learned

Simply stated, the Razorback Greenway would never have been built without the vision and generous financial support of the Walton Family Foundation. There is only one Walton family, and not every community in America is fortunate to have an engaged and dedicated private sector partner to financially support the development of local and regional greenways. Nevertheless, there are lessons learned and takeaways from the success of the Razorback Greenway that are transferable and replicable.

Trails as Transportation

When the communities of Northwest Arkansas applied for the TIGER 2 grant, they pledged that it would positively impact regional transportation and promote a culture of bicycle and pedestrian travel. Have those impacts been realized? Yes, they have. The following offers a few illustrations of this impact and transformation:

> Between 2015, when the Razorback was officially opened for public use, and 2017, annual bicycle usage in Northwest Arkansas increased by 24 percent.[5]
> The percentage of residents who now ride bikes has risen dramatically in Northwest Arkansas since the completion of the greenway. As of 2017, in Bentonville 44 percent of residents reported

riding a bike within the year, in Fayetteville 35 percent, Rogers 32 percent, and Springdale 30 percent. This is a dramatic change for Springdale, which had no bicycle facilities in 2010, and yet by 2017 a third of residents report using a bicycle.

Northwest Arkansas now has more cyclists using trails on a per capita basis than San Francisco, which cycling enthusiasts regard as one of the most bicycle-friendly cities in America.

Trail use by pedestrians increased from 2015 to 2017 by 10 percent.

Trail-user days in the region increased dramatically since the greenway opened, from 250,000 to approximately 1 million annual user days.

Approximately one-third of Northwest Arkansas residents now consider the location and availability of paved bicycle infrastructure important when deciding where to live.

One of the most dramatic beneficial impacts has occurred in Springdale. One stretch of the Razorback Greenway now connects low-income, multicultural neighborhoods to three local schools: Jones Elementary, George Elementary, and George Junior High School. Prior to the opening of the greenway, schoolchildren either rode school buses or were driven to and from school by their parents. Today, the Razorback Greenway is

Parents take turns walking students to and from local schools that are connected by the Razorback Greenway. These "walking school buses" emerged after the greenway was constructed and opened for public use. Source: Nancy Pierce photographer/Alta Planning + Design rights holder.

host to "walking school buses" in which parents take turns walking large groups of students to and from school.

Is the Razorback Greenway an effective commuter route? Yes, primarily for short distance trips in the range of two to five miles. Is long-distance or end-to-end commuting use occurring? Jeremy Pate, program officer of the Walton Family Foundation, offered a personal testimony with respect to long-distance commuting.[6] Jeremy lives in Fayetteville and bicycle commutes to his office in Bentonville, a distance of twenty-eight miles. He has noticed that during the last couple of years, the number of bicycle commuters on the trail is increasing. The ability to attract more long-distance commuters will be predicated on better understanding of the route of travel and improvements to connecting routes in which the Razorback Greenway becomes part of a larger network of bicycle-friendly travelways.

Leveraged Funding

Leveraged funding was one of the most important keys to the success of the Razorback Greenway and demonstrates the potential return on investment for other urban greenway projects. According to a study funded by the Walton Family Foundation, the investment in a regional network of bicycle infrastructure (both off-road and on-road), including the Razorback Greenway, which is the spine route of the Northwest Arkansas system, is already generating $137 million in annual economic return. A one-time investment of approximately $50 million in this network returns nearly two and half times the initial investment every year, and that ROI is expected to grow over time. The $137 million in value is composed of the following components: $25 million in bicycle tourism spending, $85 million in health-related benefits, and $20 million in bicycling expenditures by local residents.[7]

Greenways as Catalysts of Change

Throughout my thirty-five-year career planning, designing, and helping to build greenways, I have observed that these linear corridors are catalytic projects. Greenways are corridors of opportunity and have the potential to transform communities. That is certainly the result of the

The Springdale Salsa Festival is performed on the downtown portions of the Razorback Greenway and at Walter Turnbow Park. Source: Nancy Pierce photographer/Alta Planning + Design rights holder.

Razorback Greenway, which has helped transform Northwest Arkansas in ways originally envisioned in 2009 by the Walton family. The 200-acre greenway has catalyzed a conservation ethic as it weaves its way through the urban, suburban, and rural fabric of Bentonville, Rogers, Lowell, Springdale, Johnson, and Fayetteville.

Today, the two-county region can rightfully boast that it is a mecca for cycling, whether you enjoy single-track mountain bicycling, off-road paved trail cycling for fitness and health, or commuting across the region for business. In nine short years, Northwest Arkansas has been transformed from an automobile-centric community to one that embraces the culture and lifestyle of daily bicycling.

As an economic catalyst, the Razorback Greenway has helped to create new permanent jobs and supported new business development. In the TIGER grant application, the Green Team forecast that greenway development would support the creation of 350 permanent jobs within five years of opening. Most job creation occurred prior to completion of the greenway, as new bike shops, coffee houses, retail businesses, and a craft brewery opened for business in Bentonville, Springdale, Lowell, and Fayetteville. These businesses were new ventures and not relocations from other parts of the regional community, and their preferred location was,

The Razorback Greenway has spurred new business growth and development, primarily in downtown areas, like this section located in Fayetteville. Source: Nancy Pierce photographer/Alta Planning + Design rights holder.

as the design program anticipated, immediately adjacent to the greenway. The greenway is used for cultural events, including, for example, in Springdale, home of the largest Latino population in Northwest Arkansas, which celebrates an annual Salsa Festival at Turnbow Park.

Greenways and Placemaking

I firmly believe that urban greenways have become America's new "main street," where people can go on foot or by bicycle to meet each other, engage with friends, family, and even those they don't know, and share the experience of being outdoors. Charles Little, author of *Greenways for America*, summarized this well when he concluded: "To build a greenway is to build a community."[8] Greenways not only are capable of building communities; they also conserve and highlight the unique landscapes and features about communities that define a sense of place. These multiobjective landscapes capitalize on local assets, culture, and indigenous landscapes. They reveal what people love about the communities and landscapes they live in. In addition to being connectors, some greenways have the placemaking elements to become destinations unto themselves.

One of the most important takeaways of the Razorback Greenway is how it has conserved and made available to residents and tourists the unique sense of place that defines Northwest Arkansas. Celebrating the experience of place was a major objective of the trail design development program and was accomplished in part through the use of native materials, such as Arkansas field stone, the installation of indigenous plants, and Low Impact Development construction to match trail development with the flow of the land. An important design objective was to reveal the varied native landscapes and to draw on the character and authenticity of Northwest Arkansas. Reinforcing a sense of place, the greenway provides improved outdoor access to observe and learn about indigenous landscapes. Trail design mirrors the variety, rhythm, and syncopation of surrounding landscapes. In rural stretches, slow trail meanders reflect the flow line of adjacent creeks and streams. In suburban and urban stretches, trail design adapts to surrounding conditions to reinforce placemaking and promote social interaction. Trailheads, rest areas, and learning landscapes are strategically placed along the length of the trail to match the travel movements of different users.

9

White Russia

International Greenway Efforts in Belarus

Native ferns cover the forest floor of the Naliboki Forest in the Minsk region of Belarus. Photo by author.

IT WAS FIVE O'CLOCK IN THE MORNING on August 28, 2012, in the Volozhin region of Belarus, and I was wide-awake. I had not set a wake-up alarm, but there I was awake and staring at the ceiling in my room. I thought for a moment, "There is no way I am going to get out of bed, get dressed, and begin the day." Somehow, I managed to roll out of bed and began putting on layers of clothing, preparing to face a cold, dark, and dreary morning. Prior to going to bed, my Belarusian language interpreter Iryna Turouskaya approached me and informed me that our host, Anatoli Ganka, asked that I meet his father-in-law, Mihalich, at the crack of dawn to go fishing. I laughed and told Iryna that there was no way I was going to get up at that hour to go fishing—with *anyone*. She understood, agreed with me that it was something that I did not *need* to do, and then concluded that it would be a nice gesture if I would accept the invitation. I reluctantly agreed. I have no idea how my internal clock went off precisely at five o' clock in the morning, in a country that is more than 5,000 miles from my home in North Carolina, but it did, and there I was getting ready to go fishing.

I made my way down a dark staircase from my upstairs bedroom and across a small courtyard to the dining hall. It was pitch black outside, and the lights from the dining hall lit the courtyard and surrounding landscape. I could see that there were three people in the dining hall, a man and a woman seated and Mihalich standing. I thought it was a bit odd given the time of day that there would be people in the dining hall, and when I opened the door I instinctively said, "Hello," being too sleepy to remember that I was in a part of Belarus where very few people spoke English. The reply from the seated couple was equally startling, "Hello and good morning, did you sleep well?" I had never met the two who were seated but continued in English, "Thank you, yes. What brings the two of you to the Ganka homestead this time of the morning?" The young woman responded: "We just arrived from Moscow and wanted to have a morning tea." My mind was groggy and not fully functioning. "You mean you just drove all the way from Moscow, in Russia? How long a drive is that? Why did you drive all that way through the night?" I was struggling to process the conversation. "It was an eight-hour drive," she replied. "We came to hear your presentation on greenways, but we understand that first you must go fishing." At that very moment Mihalich looked at me and said, "Chai?" I thought for a moment and responded, "Yes, chai." He brought me a cup of hot black tea. Then he pointed to his watch; moments later we were sitting in his wooden boat, and he was rowing us to the far side of the pond that bordered the dining hall.

Mihalich anchored the front of the boat to the shoreline and provided me with a very long cane fishing pole that was fitted with a red and white bobber and a small hook. He fashioned rolled bread dough onto his hook and cast the line into the water. Since he could not speak English and I could not speak Belarusian, he was using hand gestures to show me how

The author and Mihalich, father-in-law of homestay owner Ganka, photographed at the family farm pond. Source: Author.

we were going to catch fish that morning. At first, I was unclear of our objective. I was observing as Mihalich was catching small fingerling fish. He was catching them with ease. I was not catching anything for the first thirty minutes on the water. Slowly I began to master the technique and was steadily bringing fish to the boat. Sitting there, fishing in silence, with someone whom I barely knew and whom I had just met briefly the day before, I had plenty of time to reflect on the situation at hand. Why was I in Belarus? What was I doing at the Ganka homestead? Why was I fishing with Mihalich? And why did a couple from Russia drive all the way from Moscow to hear my presentation on greenways?

The European Greenway Approach

The modern American greenway movement began in the 1970s when the term greenway came into popular and functional use and as communities in Raleigh, North Carolina, and Denver, Colorado, began to lay out networks of interconnected greenway corridors. Beginning in the early 1990s the American greenway concept was warmly embraced by an array of European nations. Geographically, Europe is now classified in five major regions: western, central, eastern, northern, and southern. In 1985, when the Iron Curtain was dismantled and Perestroika and Glasnost ensued, Western nations took advantage of the opportunity to engage with the citizens and countries of the former Soviet Union to establish new political and business relationships that had been either severed or suspended during fifty years of Communist rule.

Greenways in Western Europe began as a landscape typology in the late 1980s and early 1990s. Green corridors were planned, designed, and developed to support human-powered (bicycle) travel and tourism. These corridors were viewed as a way of extending tourism travel from city to countryside. Western European nations that were also part of the European Union made use of historic travel routes in the form of trails, abandoned rail corridors, and roadways to create networks of greenways. The European Greenway Association was formally established in 1998 in Namur, Belgium, as a collection of organizations that promoted the concept of greenways across the continent.

Central and Eastern European nations, including Poland, the Czech Republic, and Hungary, began greenway systems in 1990, led by a Czech-American group that promoted the idea of a Vienna (Austria) to Prague (Czech Republic) greenway. The Rockefeller Foundation provided financial and organizational support to establish the Czech Environmental Partnership Foundation (Nadace Partnerstvi) based in Brno. This partnership focused on creating a nationwide greenway program to improve the environment and support tourism. A Central and Eastern European Greenways (CEG) program emerged in 2000 and has been coordinating and promoting greenway development in Poland, the Czech Republic, Slovakia, Hungary, Romania, and Bulgaria. More recently, the CEG has begun to partner with nongovernmental organizations in Austria,

Belarus, Bosnia, Germany, Herzegovina, Macedonia, Montenegro, Serbia, Slovenia, and Ukraine.

I gained firsthand experience with the European greenway model during a 2007 visit to the Czech Republic as a guest of the Friends of Czech Greenways. Daniel Mourek, greenway coordinator with Nadace Partnerstvi, led a group of us on a guided bicycle tour through Moravia and Bohemia as part of a cultural exchange. What became clear to me during this trip were the differences in how European nations implemented greenways as opposed to their American counterparts. European greenways were primarily oriented to travel, tourism, and economic development, while for many American greenway projects the emphasis was on recreation, health, and transportation. The exchange was effective for me and my American colleagues, illustrating development techniques that we should put into practice back home. This introduction to the European point of view was important as I traveled to Belarus, a country that was in need of advice on how to diversify their economy and embrace ecotourism.

A Meeting in Budapest

It was at the July 2010 Julius Fabos International Greenway Conference in Budapest, Hungary, when I first met Valeria Klitsournova, president of Country Escape, a nongovernmental outdoors organization based in Minsk, Belarus. We were both participants and speakers at the conference. I was very impressed with her story of Belarus and fascinated by the beautiful images from her country that she shared with the audience. After my presentation concluded she invited me to visit Belarus to help her promote the benefits of greenways. Valeria, also known as Lera, had already developed a national greenway plan for Belarus. What she needed was help in convincing government officials, business owners, and citizens to support its development.

In late July 2012, Lera contacted me to share that she had secured a United States Embassy grant and was prepared to host my visit to Belarus. She asked when I was available to visit, and my initial response was "sometime in early 2013." Lera immediately responded that my timeframe was not acceptable as the grant funds were going to expire in the fall of 2012. Before I fully understood what I was committing to, I agreed to visit with her at the end of August. This put into motion a frantic effort to secure a business visa and finalize travel plans with the U.S. Embassy in Minsk. Lera was already prepared with a detailed itinerary that mapped out virtually every hour of my visit. Initially she wanted me to spend three weeks in Belarus. I reduced the travel time to twelve days.

White Russia

Translated into English, "Belarus" means "White Russia." Why it is called White Russia is subject to interpretation. There are many competing explanations. The Republic of Belarus is a landlocked Eastern European nation of approximately 10 million people (for reference, the State of North Carolina is home to more than 10 million residents). Belarus is bordered on the east by Russia, far and away the country's most important ally. Ukraine is to the south, Poland to the west, and Lithuania and Latvia to the northwest. Belarus was absorbed into the Soviet Union after World War II, and years later, after the fall of the U.S.S.R., declared its independence in 1991.

Alexander Lukashenko has served as president of Belarus since 1994 and has continued many of the Soviet era policies when it comes to state ownership of land and the economy. In 2008, U.S. and Belarusian diplomatic relations deteriorated, resulting in the expulsion of the U.S. ambassador and thirty-five embassy staff. The total number of U.S. embassy staff permitted by Belarus had been reduced to five. In the summer of 2012, prior to my visit, the Swedish ambassador had also been expelled, and the remainder of the Swedish embassy was ordered to leave the country by the end of August 2012 due to diplomatic violations that occurred in July. A small airplane, sponsored by a Swedish advertising agency, violated Belarusian airspace and was seen dropping teddy bears over the city of Minsk that contained prodemocracy propaganda.

Under new diplomatic terms, the U.S. Embassy was permitted by the Lukashenko government to fund a "speakers" program that brings American business consultants to Belarus to work with Belarusian organizations on projects that will diversify the economy. My visit to Belarus had been approved by the Lukashenko government. They knew who I was, and they knew the purpose of my visit. I worked with Carrie Lee, public affairs officer, and Elena Karpovich, staff to Ms. Lee, at the U.S.

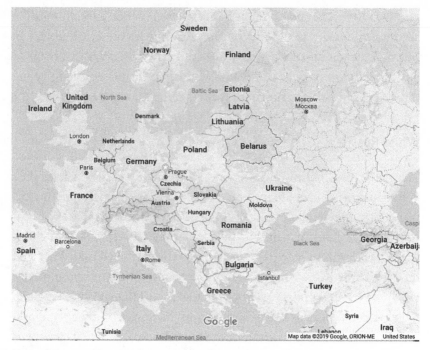

Map of Europe showing location of Belarus. Source: Google, with map data by ORION-ME.

Embassy in Minsk, to facilitate my travel. I also learned that a Belarusian language translator, Iryna Turouskaya, would accompany me throughout my visit. As I boarded my international flight to Belarus, I felt comfortable that despite the headlines and diplomatic acrimony, my visit would be successful.

Elena Karpovich of the U.S. Embassy was there to greet me at the Minsk airport baggage claim, just as she promised. There was an embassy Chevy Tahoe parked in front of the airport, and with luggage in tow we drove to my hotel in downtown Minsk. The capital city of Belarus is in many ways a throwback, architecturally, to the Soviet Union era. Minsk contains the most intact collection of Stalinist architecture from post-World War II. We drove the main street, Independence Avenue, into town. The city of Minsk was very clean, and there were reminders everywhere of the old Soviet empire. Unlike other former Soviet-era nations that have removed most of their monuments, Minsk still had original Stalin and Lenin monuments in place.

Elena shared how happy she and the U.S. Embassy staff were that I agreed to visit and work with Lera and her organization. She stressed that the work was important and that Embassy staff were grateful that I accepted the invitation to visit Belarus.

In Support of Country Escape

Later that afternoon, after I settled into my Minsk hotel room, Elena and Iryna visited with me to go through the twelve-day agenda. It was packed from early morning to late in the evening with meetings, presentations, and travel. This was going to be a grueling business trip. They bid me goodnight and told me to get a good night's rest. The next morning, Natalia Borisenko, a young, energetic staffer with Country Escape, picked me up and drove me to the offices of Country Escape, where I met with Lera and we went through the agenda again. Lera, Natalia, and I had frank and honest discussions about the opportunities and challenges of building a network of greenways throughout Belarus. They described what had already been accomplished and what it would take to implement a national strategy. The greenway vision was similar to what I had witnessed in other Central and Western European countries, with several long distance tour routes developed around a theme, such as the one in northern Belarus named "Land of the Yellow Water Lily," a 155-kilometer trail that focuses on folklore and the unique ecosystems. In fact, several pilot regions had been selected to support greenway and agrotourism development: Volozhin, Ivanova, Vileika, Rossony, Lepel, and Rogachev districts.

After our morning meeting, Lera and I headed to a rural area just outside of Minsk to visit Dudutky, a museum that she and her late husband, who was a journalist, established after Perestroika. Lera and her husband at one time owned 200 hectares (495 acres) of land that they developed into an ecotourism center. It was a working village of shops with craftsmen and other elements of rural life in Belarus. We spent time speaking with the museum staff about the greenway system and ways in which it could be linked with the museum. The museum craftsmen produced traditional goods, such as pottery, handwoven linen, and a variety of items manufactured from straw.

On our way back to Minsk, we stopped at a small village to visit a typical Belarusian homestead, a cluster of six houses. The homestead owners were incredibly gracious and happy to host Lera and me. We spent an hour and a half touring their homes and sprawling country estate, discussing the regional greenway system and ways the homestead would be linked. The six homes at the Zarechany homestead offer visitors a wide variety of activities, from outdoor BBQ to volleyball and paintball to indoor sauna and whirlpool. Their primary guests were either Russians or residents of Poland, and they were extremely interested in opening their homestead to Western visitors. With nightfall upon us, Lera bid me goodnight and loaded me into the U.S. Embassy Tahoe for the ride back to the Minsk hotel. The driver, Alexi, was assigned to my ten-day visit. He spoke no English, and we communicated primarily through gestures.

The St. Anne's Fair

On the morning of August 25, I was met at the Minsk Hotel by Alexi, Iryna, Natalia, and Lera. We drove along national highways for several hours to the town of Zelva, which annually hosts Annenskij Kirmash or St. Anne's Fair. Zelva is a rural settlement on the north end of a freshwater lake with a town population of about 3,000 and a regional population of 15,000. The St. Anne's Fair is the second-oldest rural fair (to the Leipzig Trade Fair in Germany) in Europe, dating back more than 300 years. As such it is well attended, mostly by Belarusians.

I was in Zelva to attend the fair, but more importantly to participate in a roundtable discussion hosted by local nongovernmental organizations (NGO), rural business owners, and tourism providers. I made a presentation on the relationship between greenways and fairs (or festivals). Participants wanted to know how festivals could be used to promote greenway development and how to stage community events along a regional greenway. Zelva is part of the Zelvenski Dyjarush regional greenway initiative and is located southwest of Minsk. Since the annual fair is a big draw, it was hoped that future regional greenway development would spur additional economic opportunities. Among the American examples I provided was the Grand Forks Greenway in North Dakota, which resonated with participants as Grand Forks shares the harsh cold winters and mild summers that are similar to weather in Zelva. The community of

Grand Forks has used festivals and events to promote regional tourism as part of greenway development.

Immediately on our arrival and prior to beginning the roundtable discussion in Zelva, our entire group was summoned to attend a hastily arranged meeting with the regional governor. He heard I was in town and wanted a meeting, which lasted five minutes and was for the most part a photo opportunity. Americans are as rare as hen's teeth in Belarus. I made certain that I always wore an American flag on the lapel of my blue blazer to show my pride in being an American. His one question to me was "What do you think of our country Belarus?" which was difficult to answer since I had just arrived. I provided a very short response.

After two hours of roundtable presentations and discussion, we adjourned to attend the grand opening of the fair, which was similar to festivals and fairs that I have attended in my hometown of Durham, North Carolina. Numerous vendor tents were erected within a public square. Vendors hawked everything from local food to crafts to homemade vodka. We were hosted inside one of the larger tents, where food was served and a significant amount of vodka was consumed. I soon learned that vodka was an important element of socialization throughout the country. During most days, vodka was served beginning at noon, typically as part of lunch. Consumption would continue throughout the day and into the evening. I am not a grain alcohol drinker, and during the rest of my stay I struggled to keep up with this aspect of life in Belarus.

After the opening ceremony concluded, later in the afternoon, Lera, Iryna, and I were extended an invitation to join the regional governor and other officials and farmers for an end of the harvest luncheon. It was a formal affair with many of the farmers dressed in suits and ties. There were several vodka toasts in which the farmers expressed their thanks for the bountiful harvest. As I would later learn, Belarus is the breadbasket for Russia. The Belarusian farmer's livelihood is based completely on a successful harvest. With changes in global climate, there is enormous stress on the Belarusian economy, as the harvest varies from year to year. Some years it is terrific and other years not so much. Farming is the centerpiece of the Belarusian national economy, and Russia is the most important trading partner. Therefore, one of the most important reasons for my visit was to help the national and local government agencies, NGOs, rural business owners, and tourism operators realize the potential for a national greenway system, to both diversify the economy and grow the tourism segment of the economy.

The Naliboki Forest and Native Bison

The morning of August 26 was upon me quickly. After a lengthy evening drive to Minsk and a quick dinner at the hotel, I tried my best to get a full night's sleep. The embassy car arrived precisely at 10:00 a.m., and I was asked to bring along my luggage for our trip to the Volozhin region, about a three-hour drive from Minsk. We would stay at two different bed-and-breakfast homesteads during our trip. This time the U.S. Embassy Tahoe was packed with people, as our normal traveling group was joined by two women from the United States International Development (USAID) Agency (who fortunately for me spoke English). USAID has been working in Belarus since 1999 to support activities with the hope of creating a more democratic and free-market society. One of the focus areas for USAID has been support of small business development. The USAID workers wanted to spend some quality time with me gaining insight into the potential for small business growth as a by-product of greenway development. Traveling for hours inside the U.S. Embassy car afforded us the opportunity to discuss the opportunities and challenges of small business development in Belarus. The USAID workers shared with me how challenging it was for small businesses to succeed in Belarus. Generally speaking, the borders to Belarus were closed to Westerners and the country, at the time of my visit, suffered economically from poor relations with Western nations. In contrast to this reality, I had been invited by the government and the NGO to help consider how greenways could be part of an agritourism strategy for Belarus.

Our initial stop in the Volozhin region, where we would spend our first evening, was at Alexander Bely's bed-and-breakfast homestead. After we checked into our rooms, Alexander led us on a ten-kilometer walk along the edge of the Naliboki Forest, the largest forest in Belarus, and at 5,000 square kilometers, one of the largest in Eastern Europe. The forest is home to 820 different plant species and 20 rare species of birds. The forest is culturally significant to Belarusian, Lithuanian, Polish, Russian, and Jewish culture and history. Unfortunately, for all its natural and scenic beauty, the Naliboki has a very dark history, as a place where tens

The author with representatives of NGOs in Belarus supporting agrotourism greenways at one of the entrances to the Naliboki Forest. Source: Author.

of thousands of innocent people were brutally murdered during World War II and where scores were executed during the Stalin regime. The Hollywood movie *Defiance* chronicles the Jewish resistance against the Nazis and their use of the Naliboki forest as a safe haven. In hindsight, I probably would have interpreted the forest landscape differently had I been more aware of the human tragedy that occurred.

Alexander's tour of the forest was filled with important local, national, and international cultural and natural heritage sites. There was the gutted remains of a 1930s-era mansion with its brick framework intact, burned by the Nazis during World War II. We visited an abandoned asylum, built in the late 1970s or early 1980s and vacated after the nuclear power plant, Chernobyl, exploded in 1986. It was a fascinating journey that defined the human and natural history of the forest. It was reminiscent of the 2007 greenway tour I participated in that traversed the Bohemian and Moravian regions of the Czech Republic. As with that tour, there were lots of high-value assets along the route of the Naliboki forest tour. I could easily envision a successful greenway being developed along the route that we traveled.

After lunch, we were joined by Visely, who is a manager and forest ranger in the Naliboki. Visely took us deep into the forest on a specially designed and constructed ecological walk that revealed the unique plants, animals, and ecosystems of the forest. There were several Belarusians out hunting for mushrooms as August and September are the best times to find these delicacies. Alexi schooled me on how to select mushrooms for human consumption. After our walk, we visited a memorial and outdoor Catholic Church devoted to two priests murdered by the Nazis. The memorial recognizes the sacrifice that the men made to save their congregation.

The highlight of the day occurred as we were about to leave and was a chance encounter with a herd of European bison. Visely directed Alexi to drive along the edge of the forest to an area where we might encounter the bison. But after driving for miles along forest roads, I was pretty certain that everyone had given up hope that we would see any bison. Visely explained that at one time the breed had been hunted to extinction. Currently only 1,000 of these animals live in Belarus and are difficult to locate. In fact, the majority of Belarusians have never seen the animal in its native habitat, despite the fact that it is the nation's iconic animal. But just as we were nearing the end of our search, off in the distance, grazing in an open meadow, was a herd of about forty bison. Alexi brought the Tahoe to an immediate stop, and all of us jumped out to get a glimpse of the herd. The European bison are larger than their American cousins. Even though we were a quarter of a mile away, our presence spooked them enough that the males circled the females and calves and then began to leave the meadow. I was able to photograph the herd and captured a short video before they left. To have seen the bison in person was almost a miracle.

The Ganka Homestay

Euphoric from our brief encounter with the iconic bison, we bid farewell to Visely and Alexander and headed north to Rakav, a small town on the Lithuanian border of Belarus. At one time, Rakav was one of the most important communities in the Volozhin region—a successful town with a large Jewish, Russian Orthodox, and Catholic community, all living peacefully together. Due to its location on the border of Lithuania and Poland, the town enjoyed great wealth from vigorous legal and illegal trade. We met Anatoli Ganka, a small business owner who was

Anatoli Ganka with accordion welcomes a busload of tourists to his homestay in the Volozhin region of Belarus. Photo by author.

spearheading greenway and agrotourism efforts in the region. Anatoli is a gregarious man with an infectious smile, a booming voice, and a can-do attitude. Anatoli described a time when Rakav was two-thirds Jewish prior to the arrival of the Nazis in 1939. In horrific fashion, the Nazis murdered the Jews and leveled the town. The only significant remaining structures were two churches that Iryna and I visited. Anatoli led us on a tour of a large Jewish cemetery on the edge of town that was in the process of being restored after decades of neglect during the Soviet Union era. Beyond the historic significance of the town, most alarming was the fact that there were no services in the community. There were no restaurants, no shops that sold food items, no hotels, no banks, and no retail stores. Anatoli and I discussed the negative impact that this was having on tourism. The concept of building a network of greenways across the country was challenged by the lack of support facilities that tourists depend on. Our discussion brought to light the most significant issue throughout Belarus—its struggling and failing rural economy. Outside the major urban centers, such as Minsk, there were few people and small communities that were struggling to survive economically. It appeared to me that even though it was 2012, Belarus was still suffering from the collective impact of World War II, the Stalin regime, and its membership in the Soviet Union.

After our tour of Rakav, we drove to Anatoli's homestead, which was a magnificent oasis in the midst of the undeveloped countryside. Anatoli's homestead was surrounded by large rolling hills, with spectacular and vast views across farmland, framed by groves of large hardwood trees. The homestead consisted of two large buildings, the residence and the dining hall. I was taken to one of the upstairs guest rooms, located above the family living quarters. Anatoli's entire family was engaged in the bed-and-breakfast operation (in Belarus they call them homestays). The homestay was operated by Anatoli, his wife, Larissa, daughter Paulina, son-in-law Alex, and father-in-law Mihalich. They cooked, cleaned, and maintained the homestay in a wonderful condition.

That evening, Anatoli was host to a busload of international tourists, who hailed from countries around the world. Anatoli provided seats at the dinner table for Iryna and me. I sat next to a woman from Malaysia and across from a woman who lived in Taiwan. Other tour bus participants were from Australia, New Zealand, Great Britain, and other points beyond. Anatoli and his family not only served a wonderful meal, but Anatoli provided entertainment as well, securing the services of a fabulous quartet that performed a variety of Russian, Belarusian, and American songs. The lead singer was a professional from an acclaimed theater in Minsk. Anatoli played an accordion, joining in the performance when he was not serving food or clearing dishes from the table. The entire operation was both professional and seamless. It was amazing to witness. It was after the wonderful dinner and performance that I was approached by Anatoli and extended an invitation to go fishing with his father-in-law Mihalich.

"The American Caught Lunch"

After two hours of fishing, Mihalich and I had managed to haul in more than sixty fish. The sun slowly began to make its appearance on the eastern horizon of the Volozhin region. I could see Anatoli on the opposite bank of the pond. He yelled something in Belarusian to Mihalich, who immediately scooped up both fishing poles, used the oar to push the boat off the embankment, and rowed us back to the small wooden dock. I helped Mihalich secure the boat and tackle and headed upstairs to shower and get ready for the long day ahead. When I reemerged from the lodge, I was met in the dining hall by Lera and about a dozen other visitors, who had arrived to attend the greenway seminar, including the young couple from Moscow. Lera could tell that I was a bit stressed due to my early morning fishing escapades, so she encouraged me to take my time and share the stories about my experiences with greenways. Unlike the roundtable discussion in Zelva, there was no time limit, and I was able to take the workshop participants through a variety of stories.

During a break in the meeting, I was able to more appropriately introduce myself to the young couple from Moscow. They represented an environmental NGO that was interested in pursuing greenway development. The young woman had known Lera for a couple of years, first meeting her in Moscow when Lera was visiting to promote the Belarusian greenway concept. Lera emailed her regarding my visit, and they decided to join the traveling caravan during our stay at the Ganka homestead.

We reconvened after our midday break to find that Anatoli and his family had rearranged our meeting space into a dining room and put out a feast for lunch. Anatoli emerged from the kitchen with a plate of fish in hand. He went into a lengthy explanation about the fish plate he was holding, in essence telling the group that not only did the American deliver two presentations, but also that I had joined Mihalich at the crack of dawn to catch today's lunch. The room erupted in applause.

By the time we finished lunch, Alexi had arrived and was ready to transport us back to Minsk. I posed for one more photo, this time with Anatoli and his entire family. As we were leaving, Anatoli found his accordion and serenaded me with well wishes all the way to the U.S. Embassy car. I told Anatoli that I could travel all of Belarus and never find better homestay accommodations. I thanked his entire family for their hospitality. I felt like a member of the family. They presented me with a gift of table linens made from flax—a Belarusian tradition.

The Land of the Yellow Water Lily

The next day, I was in the city of Minsk making presentations to Belarusian government agencies and other NGO representatives. This was the first formal meeting of my entire trip. One of the people in attendance was the director of tourism for the Belarusian Ministry of Sports and Tourism. Also in attendance was the director of the Belarusian national park system. Other participants were from the United Nations Development Program, USAID, U.S. Embassy, representatives from a Belarus cycling organization, professors from the state university, and a retired architect. I was most pleased that Elena Karpovich was in attendance. This was the first time I had seen her since my arrival. She had been keeping up to date with my travel and activities through phone conversations with Iryna.

My presentation focused on the economic benefits of greenways, and I could see that the message was resonating with everyone in the room. The director of tourism was taking extensive notes. One of the first questions after my presentation was in relation to cycle tracks and how they might be installed throughout Minsk. I had just finished designing the first cycle tracks in the state of Arkansas, as one element of the Northwest Arkansas Razorback Regional Greenway project, so I was able to provide a tangible example. Other questions centered on the environmental impact associated with greenway development, specifically how rare and threatened species are protected during greenway development.

After the morning workshop, we departed the city for a trip north to the "Land of the yellow water lily," one of the most scenic regions of Belarus. When we arrived in the Lepel region (Vitebsk Oblast) Lera ushered me into a municipal council chambers, and we immediately began a repeat of the morning presentations. Most of the attendees were women. There were three journalists in attendance, and a television station filmed a portion of my presentation. The questions included who was responsible for the design and development of greenway projects, referencing federal government versus nongovernmental organizations. They wanted to know what I thought of their country, and if I thought that greenways would be successful in Belarus.

After the presentation, we were invited to dinner at the home of Olga, our unofficial host in the Lepel region. I judged that Olga had to be in her seventies, and she lived in a section of the community where the homes were more than 100 years old. She offered a perspective on life in Belarus that I had not yet encountered. Most of the people that I had spent time with since my arrival were young. Olga had witnessed the

The author presents concepts for agrotourism greenways to a group of NGOs in the Minsk region of Belarus. Source: Author.

horrors and atrocities of Belarusian life from a young age through most of her adult life. She was someone worth listening to. The village of homes surrounding hers were of the same age, so I asked Olga why the Nazis did not destroy or burn her village, and her response was that the Nazis used the homes as part of their military operation, so they were spared destruction.

Olga was a big supporter of the greenway concept, firmly believing that it would help to improve the quality of life for thousands of Belarusians. Olga's home was the oldest inhabited structure I visited. Her home was made from hand-hewn wood beams. There were only two or three rooms in the entire structure. The toilet was in the kitchen, sitting in the middle of the room. Olga's husband had died years ago, but even as a widow, she was a tough, strong woman who managed the house by herself. The outside of the house was a farm in miniature with a collection of animals, fruit trees, and an extensive vegetable garden. She served an amazing assortment of fresh food from her garden. As the honored guest, I was being treated to a special meal.

We left Olga's home as the most beautiful sunset with a spectacular array of orange, red, pink, and purple colors splashed across the dark blue sky. After a forty-five-minute drive, we arrived at a Berezinsky Biosphere Reserve and checked into a lodge that was spartan but sufficient for my needs.

Berezinsky Reserve

I learned that we were staying at one of the former hunting lodges that, during the days of the Soviet Union, were reserved for the Communist elite. After breakfast, we met Natasha, our host for the rest of the day, who worked for the Berezinsky Reserve. The biosphere reserve is the oldest and only protected native landscape ecosystem in Belarus. It is a spectacular landscape that showcases the best of the Lepel region. Natasha explained that there are very strict rules in place to protect the wild animals and diverse species of plants that are unique to the biosphere. Uses of the landscape are heavily restricted. Visitors are not allowed to hunt, fish, establish temporary camps, or remove plants or animals from the biosphere. There are even limits on what can be photographed without permission. The Belarusian people and foreign visitors are not even allowed to search for prized mushrooms in the biosphere.

We toured a classroom building that also contained a large collection of stuffed animals that are representative species of the biosphere. The educational center also had extensive graphics illustrating the ecosystem components, such as the diverse species of birds and aquatic animals that inhabit the biosphere. Natasha was part of the tourism and education team at the biosphere. The management team was substantial in size and was responsible for all operations and management of the biosphere, including law enforcement, facility repair, and interpretation of the landscape. Natasha was a university graduate who specialized in tourism management.

The most important clients of the Berezinsky Reserve were the children of Belarus. Natasha shared with us that thousands of school-aged children would tour the biosphere year-round. During the summer months, nature camps were offered to accomplish more in-depth educational programs. One section of the biosphere was dedicated to an outdoor zoo, where a variety of animals were held in cages. I observed that several of the animals were in poor health, and Natasha confirmed that

A map of the Land of Water Lily Greenway system depicts the extent of the existing and planned network of trails. Source: Author.

this was a serious issue for the biosphere that her team was trying to address.

We concluded our tour with a hike into the forest of the biosphere. We visited a watchtower, fifteen meters in height, that afforded excellent views of the regional landscape. We walked along boardwalks that extended into marsh areas and contained helpful educational interpretive signage. During our hike we discussed the importance of the preserve to the regional and national greenway program. With accommodations, educational programs, and services, the biosphere could become a hub for the regional greenway system, which was similar to how the greenway system was established in the Czech Republic. Large forest preserves and cultural landscapes in Bohemia and Moravia were hubs and destinations where greenway travelers would gather and stay overnight. I suggested that Belarus could develop its national greenway system in a similar manner.

One Final Presentation

We traveled back to Minsk after leaving the biosphere. I had one final set of presentations to make in the heart of the capital city. Entering the U.S. Embassy for the final presentations of the day put into perspective the political realities of U.S. and Belarusian relations. The building where the U.S. Embassy is housed is a fortress with bomb-proof walls and doors, heavily armed security, and a thorough screening process. Even as an American in an American Embassy, I was not allowed to take my computer and backpack into the meeting rooms. My passport was confiscated and held at the guard station. I fully understood and appreciated the heightened sense of conflict that the very small five-person American staff deals with routinely each day. They are strangers in a strange land doing their very best to maintain diplomatic relations and advocate for a free and democratic society. They are constantly under surveillance and work under stressful conditions every day.

My first meeting inside the Embassy was with a group of Belarusian journalists. Iryna and I went through a brief presentation on the economic benefits of greenways. The journalists took extensive notes at a furious pace. One of the journalists, a young woman named Dasha, had an excellent command of English. All conversation was in Russian so that the others could hear questions and answers. After an intense hour of presentation and questions, Alex, the U.S. Embassy staffer in charge of the press conference, promptly ended the conversation. Dasha came up and gave me a big hug. She volunteered that she had lived on the North Carolina Outer Banks for a brief period and loved my story about how greenway development had provided an economic boost to the coastal economy.

After the press conference, I was quickly escorted upstairs for a one-on-one meeting with Carrie Lee. She enthusiastically thanked me for visiting Belarus. She shared with me that the embassy staff had been tracking all my movements, meetings, and presentations throughout the week. She emphasized that the embassy staff was very pleased with the interaction I was having with the citizens of Belarus. She mentioned that the embassy speakers program does not always produce tangible results so meaningful to the citizens of Belarus. She felt that my presentations, with a focus on practical application and economics, were what the Belarusian people needed to hear. Reflecting on our conversation, I realized how much had been accomplished over the course of ten days. I had delivered twelve separate presentations to dozens of national and local government officials, nongovernmental organizations, and interested citizens. I met and consulted with five rural businesses regarding the financial benefits of greenways and how their operations could be strengthened and linked to regional and national greenway initiatives. I conducted media interviews with a dozen national and regional television, radio, and print journalists, explaining the benefits of greenways. All these accomplishments were beneficial to Lera and her organization. It supported her decision to apply for the U.S. Embassy grant and bring me to Belarus.

My final meeting was with Ethan Goldfinch, the chargé d'affaires, the top-ranking official of the U.S. Embassy. Ethan said he had been on the job for just a few days, and he was not at all familiar with the country yet. He admitted that I had probably seen more of Belarus than he had at that time. We discussed the challenging diplomatic relations between the United States and Belarus. He reiterated what Carrie Lee had said earlier, that my presentations were particularly useful to both the U.S. Embassy and the people of Belarus, as it provided a practical program of action and focused on economic development opportunities for the rural regions of the country. He stressed that due to limitations put on the U.S. Embassy, the speakers program had become one of the primary diplomatic opportunities afforded the United States and provided an opportunity to connect directly with Belarusian people in ways that otherwise would be limited. He thanked me for agreeing to come to Belarus and share my work with the citizens and rural business owners.

Final Report and Recommendations

The last official responsibility that I had to Lera's organization, Country Escape, and with the U.S. Embassy Minsk, was to submit a report that summarized my trip and offered greenway implementation recommendations. There were several key elements of my report that are worthy of sharing. The first was my overall impression of Belarus and the Belarusian people:

Belarus is a nation with world-class natural and cultural resources. The forests, rural countryside, villages, towns and cities of Belarus provide an excellent venue for an outstanding, world-class visitor experience. One of the most important elements of this experience is interacting with the warm, friendly and hospitable citizens of Belarus. The outgoing Belarusian people are eager to open their homes, communities and resources to visitors from around the world. A program of targeted facility development and programs that support agrotourism will enable Belarus to diversify and strengthen its economy. However, this can only be achieved if Belarus makes strategic investments in agrotourism infrastructure, and when Belarus lowers the requirements for foreign visitor entry and travel.[1]

Second, among the investments that I felt that the Belarusian government and nongovernmental organizations should collaborate on, I provided a list:

1. Invest in greenway travel facilities
2. Develop better wayfinding and signage systems
3. Develop visitor lodging and facilities
4. Provide more visitor/tourist support services
5. Offer guided tours
6. Serve local food
7. Provide access to unique natural and cultural landscapes
8. Use festivals and other events to highlight the unique culture of Belarus

Finally, in my opinion, Belarus would not share in the success that other Western and Central European nations have realized from greenway development until it opens it borders to allow and encourage Western tourism. This is an enormous challenge for the government of Belarus:

The people of Belarus are the greatest asset of the country. They certainly deserve the opportunity to be rewarded for their efforts to open their communities and landscapes for agrotourism. There are limitations currently in place with regard to western tourism and travel to and within Belarus. In many respects, Belarus is an unknown quantity to citizens of the United States. However, this does not mean that Belarus is not an attractive tourist destination. More needs to be accomplished between Belarus and western nations to lower barrier to entry and travel within Belarus. It will benefit the government of Belarus to diversify its economy and support the development of rural agrotourism opportunities. Agrotourism can also provide important economic support to the citizens for these rural communities, helping to raise the standard of living and improve the quality of life.[2]

Agrotourism for American Communities

This detailed account of my visit to Belarus is with purpose. As I reflect on my trip I can't help but consider the irony of traveling to a country that has closed its borders to Western tourists and yet considers agrotourism an important strategy for diversifying its economy. My visit to Belarus was an epiphany for me. I began to wonder why states and regions of the United States aren't aggressively pursuing agrotourism strategies in a nation whose borders are relatively open and where tourists from around the world are welcomed. The Belarusian countryside and rural communities compare favorably to Midwestern communities, where I spent my teenage years. The landscapes are similar, and some of the Midwestern communities face similar challenges. There are plenty of rural communities in America that would benefit from ecotourism or agrotourism programs similar to those being proposed in Belarus. Putting aside the political differences between the two countries, the people, landscapes, ecosystems, and tourism opportunities are very similar. This is one reason I provided such an extensive account of my visit. Reconsider for a moment that my trip could have occurred in Iowa, Missouri, Nebraska, or just about any other state in the United States. What if a network of trails and greenways would be constructed to connect with fishing destinations, bed-and-breakfast establishments, heritage landscapes, festivals, hunting preserves, and environmental education centers? What would be the net benefit on the economies of Midwestern states?

In America it seems that we tend to consider the only valid job creation strategy as our ability to secure the next manufacturing facility, when in reality our economic strategies should focus on a proper assessment of local and regional resources and then attempt to match the type of business development that would best utilize those resources. One of the more successful greenway strategies of the past thirty years has been America's rails-to-trails movement, which has captured economic opportunity that rural communities can provide. Take for example the KATY Trail that bisects the middle of Missouri, and understand the significant annual economic impact that has occurred since the trail was opened. Despite these documented successes, trail-oriented development has not been warmly received and embraced by groups like the United States Chamber of Commerce and its local affiliates, nor has it been adopted in a uniform manner across the nation.

Eastern North Carolina is a great example of an economically depressed region that would benefit tremendously from a rural ecotourism or agrotourism approach. However, neither the State of North Carolina nor the federal government has spent the time and resources providing this type of assistance to declining rural areas in this part of the state. I vividly recall making a 2005 presentation to three members of North Carolina governor's cabinet, imploring them to consider making a modest $10 to $20 million investment in the development of the Neuse River Greenway in order to transform this natural resource into a destination corridor and economic engine for Eastern North Carolina. There were no takers. The economic recruitment interests in North Carolina, and to be fair many other southern states, would rather chase the next auto-manufacturing plant than take the time and energy required to figure out how they can utilize the plethora of local assets on hand to promote, market, and develop viable ecotourism and agrotourism opportunities. The truth is, there are only so many auto-manufacturing plants to chase in the world. Economic development must pursue more diversified and locally sourced opportunities.

Importance of American Outdoor Industry

To further emphasize the gap that currently exists between the aspirations of economic development and the realities that are possible, it is important to grasp the economic magnitude of America's outdoor industry. This thriving industry takes into consideration economic benefit derived from hiking, bicycling, nature travel, and other outdoor activities. A 2017 report published by the Outdoor Foundation reveals that nearly half of all Americans participate in at least one outdoor activity, equal to 144 million participants who enjoy 11 billion outdoor outings annually. The most popular outings were running, jogging, and trail running, followed by road, mountain, and BMX bicycling. Hiking was the fourth most popular activity.

The American Recreation Coalition released a 2017 Marketing Outlook report that quantifies the economic benefit of Americans' interest in the outdoors. They cited an Outdoor Industry study that concludes: "Outdoor recreation is a driving force in the American economy, generating $887 billion in annual spending and supporting 7.6 million jobs across the country." This study supports the fact that America's interest and love of outdoor recreation, travel, and tourism generates economic value that is greater than the American oil and gas services industry and pharmaceutical industry. Hiking, bicycling, and other outdoor activities represent economic opportunity that American communities should support with required infrastructure. This represents significant return on investment, business and job creation, and revenues from sales and use taxes.

Combine this information with the fact that travel and tourism is ranked as either the first, second, or third largest economic engine in all fifty states.[3] Think about that for one moment, because I am almost certain that most who will read this had no prior working knowledge of this economic reality. To further illustrate this gap in knowledge, consider that in 2015, domestic travel in North Carolina generated $21.9 billion in revenue, which was a 3 percent increase from the annual revenue in 2014. North Carolina is the sixth most visited state in the nation. In 2015, approximately 54.6 million people visited the state of North Carolina and enjoyed a wide variety of outdoor resources, from the beaches to the Piedmont region and the mountains. Visitors to the Old North State spent on average $55 million per day, generating $2.9 billion in state and local tax revenue. And there is more. Almost one in every ten employed North Carolinian works in the travel and tourism industry, totaling 211,490 jobs.[4] And yet, with all of the good news listed in this paragraph, travel and tourism and the opportunity that it holds to transform and fuel the American economy remains for the most part under the radar and unknown not only to the average American but also to the average business person in America. I know this for a fact because I have made presentations in several states citing similar statistics for Georgia, Florida, South Carolina, Tennessee, and Missouri, to name a few, and the vast majority of people that I speak with tell me that prior to my presentation they had no understanding of the economic importance associated with travel and tourism.

Greenway Tour Routes

American greenway systems can and should be developed to support cultural and natural heritage tours. This is an area of work that needs additional focus and effort by greenway planners, designers, and operators to accomplish this objective. Not all constructed American greenways would achieve these objectives, but there are a number of existing long-distance greenways that have much to offer in terms of natural and cultural heritage attractions. They should be operated in a manner that caters to hiking and cycling tourists who want to better understand the culture and natural history that has shaped different regions of the United States.

In the Czech Republic, the Friends of Czech Greenways constructed and mapped long-distance bicycle-touring routes that are publicized in printed brochures and on websites. Some of these routes offer guided tours; however, adventure seekers can also complete this travel on their own. The Friends work with hotels, restaurants, and other businesses through a certification process to ensure that they are willing to accommodate the needs of bicycle tourists. This includes places to park and store bikes, a wash rack or equivalent to clean equipment, restaurants that can accommodate casually dressed customers, and other similar accommodations. Again, why are Americans not pursuing similar programs of accommodation throughout the United States?

The Culture of Bicycle Tourism

In the United States, Adventure Cycling is one example of a tour operator that makes available 47,000 miles of bicycle routes for travel and tourism. Outside the offerings of Adventure Cycling, bicycle tourism is not a popular pursuit or in some cases even worth the risk of life and limb. In many parts of America there is hostility toward bicyclists, and much of it is unwarranted. The United States is an automobile-oriented culture and society, and some in our nation see a cyclist riding in the road as an opportunity to inflict pain and bodily injury. There are far too many stories throughout America of bicycle clubs that have lost valued members and successful community citizens to carelessness or acts of intentional or unintentional death by motor vehicle.

In Europe, the culture is for the most part very different. At least that has been my experience. Cyclists and motorists often share the same roadway, and most of the time motorists yield the right-of-way to the cyclist in order to avoid travel conflict. America needs to embrace the culture of cycling and bicycle tourism. It benefits all of American society, producing healthier people, more economic opportunity, and fewer impacts to air, water, and land.

A targeted rural economic development strategy that features the development of interconnected greenways would yield positive results for many American communities. We should not have to travel to foreign countries to fully understand and appreciate the value that this type of coordinated federal, state, and local economic strategy can and will have on American communities.

10

America's Longest Urban Greenway

East Coast Greenway, from Maine to Florida

Marketing poster for the East Coast Greenway. Source: East Coast Greenway Alliance.

"WHERE ARE YOU GOING?" a young boy cried out as I steered my hybrid bicycle past rows of residential homes in southwest Philadelphia. "To New York City," I responded. Running alongside my bike, he continued: "On a bicycle? Are you crazy?" Laughing, I replied: "No, not at all, go get your bike and ride with us." He stopped running and, with his voice fading in the distance, responded: "I have to ask my mom for permission."

Similar exchanges occurred regularly during the fall 2008 "Chairman's Ride" to raise awareness and funding for the East Coast Greenway. I was leading a group of fifteen bicycle riders on a 225-mile, seven-day tour from Wilmington, Delaware, to the "Big Apple" (New York City). Admittedly, we were an unusual-looking group of riders: three were older than seventy-five, and at forty-nine years old I was one of the younger participants. We were colorfully dressed in a variety of clothing, from cargo pants to spandex, with some of us sporting a green, blue, and white East Coast Greenway jersey. Several bicycles were equipped with large white flags mounted on tall fiberglass poles emblazoned with the East Coast Greenway logo. Collectively, our group looked like we could be participants in a holiday parade. The ride was one component of the annual East Coast Greenway Alliance (ECGA)–sponsored "Close the Gaps" campaign. Our journey began in the heart of Wilmington, Delaware, on the banks of the Christina River, and concluded in New York City, on the banks of the Hudson River. We rode less than 10 percent of the East Coast Greenway, but to a person, it was one of the best bike rides most of us had ever experienced.

The Urban Appalachian Trail

The East Coast Greenway is envisioned as a 3,000-mile mostly off-road trail extending from the Maine-Canadian border south to Key West, Florida. To date, about one-third of the greenway, approximately 1,000 miles, is completed off-road trail. The remaining 2,000 miles is on-road, defined with specially designed ECGA road signs and referred to as the interim route. In the coming decades more of the interim route will transition to permanent off-road trail. Despite its daunting length, in reality the East Coast Greenway is a collection of local, regional, and state trails that will eventually be joined together across fifteen states and the District of Columbia, passing through 450 cities and towns, along America's eastern seaboard states.

The greenway is currently the nation's longest connected urban-oriented bicycling and walking route and is often referred to as the urban version of the Appalachian Trail. The greenway is routed through the center of America's largest cities: Boston, New York, Philadelphia, Baltimore, Washington, DC, Durham, Charleston, Savannah, and Miami. The corridor is located within a four-hour car drive of approximately 120 million people and supports human-powered travel for a diverse array of users, including bicyclists, walkers, horseback riders, in-line skaters,

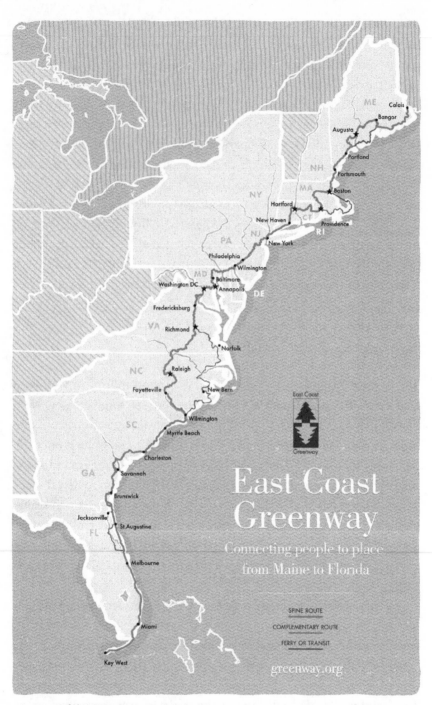

Map of the East Coast Greenway. Source: East Coast Greenway Alliance.

runners, cross-country skiers, and wheelchair users. Some sections of the greenway cross rivers where travel is accommodated by ferry boat, canoe, kayak, or other vessel.

The East Coast Greenway Alliance, headquartered in Durham, North Carolina, is the nonprofit organization responsible for stewarding the vision and championing development of the trail. The alliance estimates that approximately 11 million people use the greenway every year.[1] The long-term goal of the alliance is to increase annual use to 50 million during the next two decades, working with local trail partners along the route to close gaps in the trail network and open off-road trails that link one segment to the next, across cities, metro regions, intrastate, and interstate, north to south, from Maine to Florida. The alliance takes pride in the fact that the East Coast Greenway addresses relevant urban community issues, such as:

1. *Active and Healthy Lifestyle*: The East Coast Greenway provides people from all walks of life with no-cost access to a high-quality outdoor environment where they can participate in an active lifestyle close to home, school, and the workplace.

2. *Ecotourism*: The greenway has already become a sought-after destination experience. Connected long-distance off-road segments are safe and accessible for families. Using digital online tools provided by the alliance, users can create their own travel itineraries, as you will see later in this chapter. Greenway users can visit some of the nation's most popular national landmarks and historic landscapes. Travel is accomplished by bicycling or walking, and the greenway is linked to metropolitan and national transit systems (such as Amtrak). Since the greenway is found mostly within urban or suburban landscapes, travelers can easily find lodging at a variety of hotels, historic inns, and bed-and-breakfast locations.

3. *Alternative Transportation*: The greenway provides a corridor that supports a nonpolluting, human-powered alternative to automobile commuting within large metro areas. The greenway weaves through densely populated areas and links to some of the most popular local, regional, and national destinations, including

employment centers, shopping centers, city center, sports venues, schools, and college campuses.

4. *Economic Development*: The greenway provides substantial return on investment. Economic studies of the greenway reveal that it improves the value of adjacent properties, attracts tourists, supports a diverse workforce, creates opportunities for new business development, and in general improves the economic well-being of host communities. A 2017 study of the seventy-mile greenway route in North Carolina's Research Triangle region reveals $90 million in annual economic benefit based on the following factors: annual commuter miles bicycled, miles walked per year, hours of physical activity per year, temporary and permanent jobs created, annual health and wellness benefits, and one time property benefits.[2] Again, this is for just one seventy-mile segment of the greenway. Imagine multiplying similar benefits across more than 1,000 miles of completed greenway.

A Point of Light

I first learned about the East Coast Greenway in the winter of 1992 during a meeting at the White House. I was working with Anne Lusk, one of President George H. W. Bush's Thousand Points of Light, so honored for her work establishing the Stowe Recreation Path in Vermont. Anne was then, and is to this day, a bundle of energy and passion dedicated to researching and understanding factors that influence the design and development of safe, inclusive, healthy, affordable, and enjoyable trails.

For the past decade Anne has worked as a research scientist at Harvard University's T. H. Chan School of Public Health. Study of public health was not her original career path. In 1971, Anne earned a diploma in fashion design from the Les Écoles de la Chambre Syndicale de la Couture Parisienne, Paris, France. She was working for Geoffrey Beene in New York City when in 1972, during a ski vacation to Stowe, Vermont, she met ski instructor and eventual husband Charlie Lusk. While living in Stowe, Anne pursued community preservation as an outgrowth of her 1975 University of Vermont Masters in Education and Historic Preservation and became a founding member of Historic Stowe Inc. Anne led preservation efforts throughout Stowe, first saving the old Stowe High School from demolition and then helping to designate 124 Stowe homes on the National Register for Historic Places. In 1981, her interests changed, and she was hired by the Town of Stowe as the community bike path coordinator, which led her into the design and construction of the 5.3-mile Stowe Recreation Path. Anne overcame a myriad of challenges involved with the project and during a stretch of seven years led a communitywide effort to build the off-road trail. In recognition of her work, and the Town of Stowe's achievements in building the path, Anne and the Stowe Recreation Path were designated the 119th Point of Light by President Bush.

Anne's experience with the Stowe Recreation Path ignited a desire to learn more about the benefits of community-oriented trails and greenways. We met for the first time in October 1988 at the Unicoi State Park and Lodge in Helen, Georgia, the night that the American Trails Network and the National Trails Council merged to form American Trails, Inc. I was elected board chair of the new organization. Anne was attending to learn about and become better connected with the people who were part of the emerging American greenway movement.

Several years later, our paths crossed again. In February 1991, Anne asked if I would accompany her to meetings with the Bush White House, Office of Domestic Policy. The goal of the meetings was to establish a national policy framework for greenways. The Office of Domestic Policy was intrigued by the potential of greenways, and Anne, by virtue of her Point of Light status, had made a solid connection with Ed Goldstein in Domestic Policy. Our meetings with the White House occurred almost monthly throughout 1991 and into the spring of 1992. It was during one of our monthly White House meetings that Anne told me that she was working with a group of people to form a new organization and complete an inaugural bicycle ride in support of a greenway that would link Boston, Massachusetts, to Washington, DC.

The early 1990s were important to the growth of the American greenway movement. A national vision for an interconnected network of greenways was emerging from different quarters across the nation. The national interest in interconnected networks of greenways got a significant boost in 1985 when President Ronald Reagan appointed the President's Commission on Americans Outdoors (PCAO) to examine ways in which Americans were using the outdoors for travel, tourism, recreation, health, and fitness. One of the commission's recommendations supported the development of a national network of greenways: "The Commission envisions a new system of 'greenways' along scenic highways, waterways and trails, linking communities and cities and the expansion of the federal estate."[3]

When the commission released its report in 1987, it was immediately enveloped in controversy as some interpreted the objective and recommendations of the report as supporting a federal land grab of private property. President Reagan did not support the findings of his own commission, and when confronted with the controversy he directed the office of Vice President Bush to issue a formal written response. Island Press published the original unedited version of the PCAO report and recommendations, and the Vice President published a rebuttal.

At the same time as the PCAO report was swirling in controversy, I was in the process of starting my consulting company Greenways Incorporated, which opened its doors for business in August 1986. Later that same year, the Rails-to-Trails Conservancy was established under the leadership of David Burwell and Peter Harnik, devoting energies to transforming abandoned rail corridors into trails. American Trails, Inc. was launched in 1988, and by the summer of 1990 a coalition of trail and greenway advocates led by Dave Startzell, chair of the National Trails Agenda Task Force, and Bill Spitzer of the National Park Service, Recreation Resources Assistance program, published a report titled *Trails for All Americans*, which recommended building a national network of trails that could be accessed and used within fifteen minutes of where citizens lived and worked. Trails and greenways were on the minds of many Americans. Energy and enthusiasm were emerging from across the nation in support of planning, designing, and building a network of interconnected trails and greenways.

Birth of an Alliance

Patricia King, of Newton, Massachusetts, had a vision for a long-distance greenway that she hoped would span east coast to west coast. Pat was born in Richmond, Virginia, and raised in Pennsylvania. She attended

Westhampton College (the University of Richmond, Virginia, women's college) majoring in history. Influenced by the groundswell of interest in the environment, after graduation Pat moved back to Pennsylvania to work with the Chester County Horticultural Society and soon thereafter became the secretary of the first Earth Day Committee in Philadelphia. Moving to Massachusetts in the mid-1970s, she took a deeper interest in environmental issues, becoming the chief lobbyist at the Environmental League of Massachusetts and later director of the Department of Communications and Legislative Affairs for the league. By the 1980s she was employed with the State of Massachusetts Department of Environmental Protection and began to understand the role that bicycle transportation played in environmental policy. One of the limitations she immediately noticed was that most bicycle pathways were short and not connected to the larger community framework, much less larger ecosystems.

In the spring of 1991, Pat crafted a proposal titled "Planning and Implementation of a National Bicycle Touring Trail." Pat imagined a greenway that would link together national parks from Maine to California. Armed with her carefully written proposal, she traveled to Washington, DC, in hopes of meeting like-minded individuals with whom she would share her vision. She met David Burwell and Peter Harnik at the offices of their newly formed Rails-to-Trails Conservancy organization. Later in the day, she met with John Fegan, bicycle program manager for the U.S. Department of Transportation, Federal Highway Administration, who advised her that most volunteers would be more satisfied investing their energies in local and not national long-distance trail projects. She finished the day meeting with Andy Clarke, vice president of the Bicycle Federation of America. After a long day of meetings, Pat was disappointed that no one shared her vision for a coast-to-coast trail linking the national parks. It was viewed as too ambitious at the time. She had no interest in building a five-mile trail in her hometown; her interest was in spearheading a long-distance trail, linking significant landscapes across the nation.

One of John Fegan's suggestions resonated with Pat—that she should attend an upcoming bicycle conference at the Massachusetts Institute of Technology (MIT) in Cambridge to present her idea and proposal. She attended the conference, where she met Dave Lutz, executive director of the Neighborhood Open Space Coalition in New York City, and Karen

Votava, director of open space planning for the New York City Planning Department, both of whom went on to work with Pat to launch the East Coast Greenway.

On the Friday after Thanksgiving in 1991, a nine-person group of greenway advocates, led by Pat King, gathered at the AYH Amsterdam Hostel in New York City to continue the discussion begun at the MIT conference—establishing a new long-distance greenway between New York and Philadelphia. The group included King, Lutz, Votava, Jessica Mink, Aldo Gherin, Tom Pendleton, Danny O'Brien, and Ken Withers. As the group met throughout that late November day, they dreamed and debated the merits of a long-distance trail. The more they talked, the longer the travel route became until it extended from the Boston Common, through the heart of New York City, past the Liberty Bell in Philadelphia, and all the way to the National Mall in Washington, DC. It was an ambitious vision, particularly given that the route would travel through the heart of America's largest and most densely developed cities.

Pat King wasn't the only greenway founder with big dreams. Karen Votava wanted to be involved in a big, bold, and impactful greenway-oriented initiative. A 1965 graduate of Douglas College (now part of Rutgers University), Karen worked as an intern with the New York City Planning Department, which steered her in the direction of earning her 1970 master's degree in city planning from Yale University. After graduation she returned to the City of New York, where over time she rose through the Planning Department ranks and found herself in charge of open space and park planning. Through this work she met Dave Lutz at the Neighborhood Open Space Coalition and began to expand on one of the modern expressions of greenway development in the city—the Brooklyn-Queens Greenway. Karen went on from that project to serve as the lead author for the New York City Greenway Plan, which extended across the five boroughs of the city, spanning more than 350 miles in length and linking city parks with neighborhoods and popular destinations. "We snuck plan approval through Mayor Dinkins office," Karen recalls. Today, much of that plan has been implemented and has helped make New York City more walkable, bikeable, and livable.

Karen envisioned the potential for interconnected regional greenways through America's northeast metropolitan region. Now she joined with Pat King, Dave Lutz, and others in considering the merits of a long-distance greenway that would link New York City to metro areas north and south. After she completed the NYC Greenway Plan in 1993, Karen left the City Planning Department in 1994 and moved to Wakefield, Rhode Island, with her husband Bob Votava to pursue the East Coast Greenway. In 1996, she was hired in a part-time capacity as the first executive director of the ECGA.

Another contributor to the early success of the East Coast Greenway was Bill O'Neill, who at the time was a partner and owner of Fuss and O'Neill, a civil engineering firm headquartered in Hartford, Connecticut. Bill was a gubernatorial appointee to the Connecticut Greenway Council when in the winter of 1992 he received a request from Anne Lusk, who was in search of sponsors and riders for the inaugural East Coast Greenway ride. Bill volunteered that his firm would help prepare route maps. He also volunteered that two of his four children, Steve and Kathleen, would be trail riders. Bill and his wife, Carol, helped out by driving the sag wagon during part of the journey. After the ride was completed, Bill joined the Board of Trustees, became ECGA Connecticut State Committee chairman, and his firm assisted the alliance as corporate sponsors preparing state-by-state illustrative maps depicting the greenway route.

The greenway was a concept and a glorified idea that needed grounding when in July 1992 a group of ten riders departed Boston, Massachusetts, heading south to Washington, DC. The riders included Pat King, Anne Lusk, her daughter Katharine Lusk, Katharine's friend Alexis LaRow, Bill O'Neill's children Kathleen and Steve, and William W. Johnson from Washington, DC, who like Anne Lusk was a Point of Light. William Johnson brought along his son Lamont and friend Anthony Williams. The goal of the ride was to evaluate the feasibility of the idea for an East Coast Greenway through a marketing and promotion event that included frequent stops for interviews with local media. It was an ambitious journey of 600 miles to be completed within a thirty-day timeframe, much of which would have to be navigated using roads, as there were very few completed off-road greenways within the defined travel corridor.

The ride almost never occurred due to difficulties in securing enough riders and working out the logistics of the trip. Anne Lusk, Pat King, and Bill O'Neill worked hard to overcome these challenges—raising funds, securing Cannondale bicycles (pro bono) and bicycle accessories (rain

Inaugural riders on the East Coast Greenway, circa 1992. Source: East Coast Greenway Alliance.

gear, helmets, locks, and bike bags) for each of the riders, reserving lodging, and making certain there was ample food for the entire trip. Anne printed "East Coast Greenway" T-shirts for each of the riders. The ride was a success and galvanized the group and their efforts. It proved to the visionary founders of the East Coast Greenway that while the journey was difficult, it was also practical and possible to accomplish.

An Alliance of Local Partners

By the fall of 1992, enough people had shown interest and enthusiasm in the trail to form a not-for-profit organization, which would serve as champion of the greenway vision. Under the leadership of Pat King, the group formed the East Coast Greenway Alliance. From 1992 to 1995, the fledgling organization began to slowly ramp up its activities, securing a website (www.greenway.org), filing articles of incorporation, completing a strategic plan, securing trademark protection, and adopting bylaws. Looking back, the founders now admit that they were in over their heads. None of them had ever formed such an organization prior to setting up the ECGA. Collectively they had little working knowledge of how to run a nonprofit. It was definitely on-the-job training for virtually all involved.

The alliance continued to sputter along but was bailed out by its partners, all of whom were excited about the greenway vision. In 1993, the National Park Service agreed to assist in drafting a new strategic plan in partnership with the alliance. By the close of 1995, the alliance had been incorporated in New York and established official headquarters in Wakefield, Rhode Island. The following year, Karen Votava became the executive director, and in 1998 a board of directors was seated with Pat King serving as inaugural chair.

The ECGA emerged through the years as a strategic organization capable of partnering with local, state, regional, and federal governments, private sector consultants, businesses, landowners, and enthusiastic volunteers to bring the East Coast Greenway to life. The genius of the alliance is that it owns nothing: not a piece of land, no construction equipment, no maintenance equipment, not even one foot of completed East Coast Greenway trail. The alliance is the keeper of the flame, steward of the vision, and champion of getting a 3,000-mile trail completed and connected. In the mid-1990s, the alliance established guidelines for the types of trails that they felt should comprise the route of the East Coast Greenway. Those guidelines define trail tread surface, trail width, ability to satisfy national trail design standards, and the capability to operate and maintain the trail. The alliance executes an agreement (memorandum of understanding) with local, regional, state, and federal partners, who are asked to meet the development and operational standards so their local greenway is formally accepted and adopted as part of the branded East Coast Greenway. For example, one of the most interconnected off-road sections of the greenway is located in the Raleigh-Durham metropolitan region and is called the Cross Triangle Greenway. This section of greenway is composed of six separate locally funded, constructed, and maintained trails. The Cross Triangle Greenway is one of the longest continuous off-road greenways along the entire 3,000-mile route. The alliance did not build one foot of this stretch of greenway, but it did certify and accept this section as part of the official trail route.

A Legacy of Leadership

As chair of the inaugural board, Pat King worked hard over the course of nine years to birth the alliance and formalize its organizational structure.

By the close of 2000, she was ready to take a break from the alliance and from her job at the State of Massachusetts. She set off to earn an art degree and enjoy some freedom from the responsibilities of work and life. Dave Dionne, chief of trails and greenways at Anne Arundel County, Maryland, was tapped as the next board chair. Dave first met Karen and members of the ECGA team in 1994 when they were walking along the Baltimore and Annapolis Trail scouting new southern routes in Maryland. He immediately understood the power of the vision and the opportunities for success and joined the alliance board in May of 1995. In 2001, when Dave became board chair of the alliance, he was already an experienced trustee and was familiar with the ECGA operations. Dave patiently endured the chaos that is often associated with start-up nonprofits. He agreed to serve as chair and set about making his mark on the organization. First and foremost on Dave's mind was that the organization needed to conduct its business in accordance with professional standards. That meant that meetings had written and printed agendas, board of trustees discussion was to be limited to the topic at hand, and trustee meetings should never last more than four hours. During Dave's tenure, and with Karen growing more accustomed to her role as executive director, the ten-year-old organization leapt from its adolescence phase and began to flourish. Dave's term as chair of the board ended in 2004 when personal and professional obligations led him in a different direction.

My formal involvement in the East Coast Greenway Alliance began in in 1996 when Karen Votava contacted me and asked if I could serve as the ECGA North Carolina State Trail Coordinator. I declined due to my travel schedule and workload. I asked Tricia Tripp, one of my staff, if she had time and would be willing to serve. She agreed and served as state trail coordinator in 1996. The alliance was beginning to make a push into southern states and scheduled a workshop in Pinopolis, South Carolina, where five southern states, Virginia, North Carolina, South Carolina, Georgia, and Florida, met to discuss the greenway route and alignment. Tricia attended the meeting and returned to the office with tremendous enthusiasm for the project. She also announced that she was getting married and could no longer serve as the state trail coordinator. I assumed those duties in 1997. Karen encouraged me to attend meetings of the alliance as North Carolina's representative, and in 1999 I made time in my schedule to attend my first alliance meeting, where I was asked to join the board as a trustee. I served in that capacity for the next twelve years.

When Dave Dionne announced he was stepping down as board chair, my phone began to ring. Karen Votava painted a bullseye on my forehead and wanted me to serve as the new board chair. As I recall, I said no to becoming the new board chair nine times, and by the tenth call, I relented and became chair in 2005. During the next five years, I led the alliance through a major reorganization and, when Karen announced her retirement in 2009, facilitated a transition in alliance leadership. I led the search committee that eventually hired Dennis Markatos-Soriano, who has served as the alliance executive director since 2010.

Pick a Day, Three Days, or a Week

The questions that I have heard most frequently regarding the East Coast Greenway include: Why should we create long-distance trails in the first place? Why do we care that a 3,000-mile-long trail exists along the eastern seaboard states of America? These are valid questions. Historically, long-distance trails made sense when tens of thousands of Americans were using them to migrate across the continent. But there we are talking about American history of the seventeenth, eighteenth and nineteenth centuries. The United States was settled by risk-taking pioneer men, women, and children who fanned out from east to west along rugged wilderness trails. Our nation today is a product of those pioneers and the cross-country trails they created. The East Coast Greenway does not retrace a specific historic travel route. The greenway does follow parts of the Old Philadelphia Wagon road, the major north-south highway of Colonial America; however, the greenway was not envisioned and created for the sole purpose of retracing that road or any other historic travel route. Why then do we need to create an interconnected, 3,000-mile-long modern pathway through fifteen states?

Having met with and discussed this particular issue with hundreds of greenway users, elected officials in towns that are connected by the greenway, and with business men and women who envision future economic success from the greenway, the simple answer and explanation is people love to be connected to big, transformative ideas and concepts. They like knowing they are part of something bigger. When asked, "Would you ride the East Coast Greenway from end to end, the entire 3,000 miles?" most people respond: "No way." But most love the idea of being located on the route and part of a project that puts their community on the map and is situated along a corridor that stretches from Maine to Florida.

I can attest to the impact that the East Coast Greenway has had on the communities through which it is routed. My hometown of Durham, North Carolina, not only takes pride in having the greenway pass through the community, its development helped revitalize the downtown. It is fun to contemplate that as the greenway becomes a reality on any given day I can literally grab my bike (I live two miles from the greenway) and pedal north or south for as long as I want, all the way to Maine or Florida if I have the desire to do so. This is no ordinary greenway; it is extraordinary, and the route of the trail connects me to landscapes and places that are interesting and fun to visit. I think that all of us are intrigued by the opportunity to bicycle or walk along a pathway that is off and away from roads, that meanders over hill and dale, alongside streams, through thick forests, through the heart of cities and towns, and across varying landscapes. That is the lure of adventure and of travel, and the East Coast Greenway affords that opportunity.

To date, fewer than twenty people have ridden the entire greenway from end to end. The vast majority of the 11 million annual uses of the East Coast Greenway occur as daily commutes to work or school, for exercise, or for some other activity such as shopping.

The great thing about the East Coast Greenway is you don't have to be an end-to-end rider to enjoy its outstanding attributes. All you need to do is pick a distance, what is the most comfortable length of a bicycle ride or walk, choose your desired start and stop points, and enjoy the journey. I am not a hard-core cyclist. I am not one of those weekend century or half century riders. I mean no offense to those who are and who enjoy the riding clubs that exist all across America. My point is, you don't have to be an extraordinary bicyclist to enjoy the East Coast Greenway. That is what makes the greenway such a perfect venue. It caters to a wide range of skill sets. For me, twenty-five to thirty miles on any given ride is enough. I like to go slower, enjoy the scenery, eat a hearty lunch, visit historic sites, and be more of a tourist than a rider.

To illustrate how anyone might use the East Coast Greenway, I will share three separate journeys I participated in along the trail—excursions that lasted one day, three days, and seven days. I present these chronologically, in the order in which I experienced the greenway.

A Three-Day Ride: Baltimore to Washington, DC

A long weekend trip might be one way for you to experience and enjoy the benefits of the East Coast Greenway. There are sections of the East Coast Greenway that are perfectly suited to a three-day ride and can be accomplished by riders who have some experience with bicycle riding. I took my first long-distance greenway trip in June 2007, riding for three days from Baltimore to Washington, DC. This segment of the East Coast Greenway winds its way through industrial areas, along waterfronts and rivers, through the outskirts of two large American cities, and into the heart of our nation's capital.

It began in December 2006 during a brief conversation with my friend and colleague Mark Fenton, who at the time was vice chair of ECGA Board of Trustees. Mark and I had worked together on the Grand Canyon Greenway project, joined the ECGA board about the same time, and were considering bicycling a portion of the greenway. Shortly after we confirmed our commitment to bike a portion of the greenway, we contacted two other partners in crime, Jeff Olson, who was then a member of the ECGA National Advisory Board, and Bob Searns, who was chair of the Board of Trustees for American Trails. We invited them to join us on the ride, and they immediately and enthusiastically accepted. We all quickly figured out that bicycling the entire 3,000-mile East Coast Greenway was out of the question, as none of us had the time available to accomplish an end-to-end ride. What I found especially appealing about this route of the greenway was how close it was to millions of people, who would find the trail easy to access and use for a similar three-day adventure. After consulting with Karen Votava, the four of us decided to focus our efforts on a 100-mile segment of nearly complete greenway from Baltimore's Inner Harbor to Union Station in Washington, DC.

On a sultry summer morning, June 13, 2007, our small group of riders began a journey that clearly illustrated to all of us just how important and meaningful the East Coast Greenway can become. Joining us on the ride was Greg Hinchcliffe, an American Airlines pilot, who was chair of the ECGA Maryland State Committee; Michael Olivia, the ECGA Mid-Atlantic Region trail coordinator; Bob Searns's wife, Sally; and Elizabeth Brody, an ECGA trustee from New Jersey. Beth Brody deserves special mention as she was at that time an iron-willed and fit 70-year-old cyclist who was excited to participate in the ride and continued for many years after to participate in other ECGA-sponsored rides. Beth joined the board of trustees in 2005 and became the ECGA New York State Committee chair in 2006. Most important, Beth has been one of the most important philanthropic supporters of the greenway. Her contributions to the success of the greenway as trustee, funder, and mentor are legendary.

The goal of our ride was simple—to help the ECGA staff identify gaps in the 100-mile-long stretch of greenway and determine how best to close the gaps and complete this stretch of the greenway as an off-road experience. All the participants wanted to demonstrate that this 100-mile ride could be a meaningful way for average Americans to become more active in their daily lives. For those who might travel this route for commuting, recreation, or tourism, the ride represented an affordable alternative to driving a car.

We began the ride with a "gap closing" ceremony at the Inner Harbor, by unveiling new ECG signage that provides direction of travel information for greenway users through downtown Baltimore. Routing the greenway through the heart of big cities is a challenge, and we wanted to make certain that the signage was visible and easy to follow.

Throughout our Maryland journey, the group was supported by former alliance board of trustees chair Dave Dionne, who at the time was superintendent of the Anne Arundel County Parks and Recreation Department. Dave and his staff did a tremendous job making us comfortable, providing us with rental bikes, returning the rental bikes upon completion of the ride, and feeding us.

The first day of our journey included a thirty-five-mile ride from downtown Baltimore to the edge of Annapolis. Greg Hinchcliffe served as our guide out of Baltimore along newly opened sections of the Gwynn Falls Trail. We connected to the Baltimore-Washington International Airport Trail, which encircles the airport, and then connected to the historic Baltimore and Annapolis rail-trail. We stopped briefly for lunch to enjoy a

Maryland delicacy, jumbo crab cakes. As we approached Annapolis, we realized that this journey was so easy that anyone could tackle and enjoy it. We concluded our day with a dinner hosted by the Friends of Anne Arundel County Trails at Jonas Green Park, overlooking Annapolis harbor. The park has the great distinction of being the intersecting point of two long-distance national trails, the East Coast Greenway and the east coast to west coast American Discovery Trail.

Day two began bright and sunny. We were greeted at City Hall by Mayor Ellen Moyer, an enthusiastic supporter of the East Coast Greenway. Mayor Moyer served on our National Advisory Board at the time, and declared June 14, 2007, East Coast Greenway Day. Her chief of staff, Steve Carr, provided a three-hour guided bicycle tour of Annapolis. We traced the route of the East Coast Greenway from downtown through historic parts of the city, concluding at the U.S. Naval Academy football stadium. The tour was one of the highlights of our ride. Annapolis is steeped in history and significance and is a must-see community along the greenway.

After our tour of Annapolis, Dave Dionne and crew loaded our bicycles into support vehicles, and we traveled by car approximately ten miles west of Annapolis to the headquarters of Anne Arundel County Parks and Recreation. Along the way, Dave drove us along the route of a proposed county trail, portions of which were developed in 2008 and opened for public use. Another gap in the greenway was on its way to being completed in Maryland. Dave and his staff treated us to a good old-fashioned summer picnic of hotdogs and hamburgers at park headquarters.

After lunch we were again shuttled by vehicle to Bowie, Maryland, to meet with Bob Patten, state committee chair for the District of Columbia. Bob arranged for several local cyclists to lead us on a tour of the Bowie to College Park section of the greenway. Our goal was to evaluate three separate routes that were under consideration by the DC and Maryland state committees and to help them determine which route was the most feasible. Our small group was divided into three teams. Bob Patten supplied us with local, knowledgeable cyclists to lead the way. Sally Searns and Beth Brody cycled the northernmost route, which traversed rural agricultural lands. Bob Searns and I rode the middle option, which used highways and local roads in Greenbelt, Maryland. Jeff Olson, Mark Fenton, and Mike Olivia traversed the southern leg, down the full length of the Washington, Baltimore, and Annapolis Trail and through College Park.

After we completed our separate rides, we reunited at a local hotel in College Park and headed to the REI Store in College Park, where we were joined by several bicycle enthusiasts to discuss the merits of each route. There was a unanimous opinion that the northernmost route, through the agricultural lands, was the most desirable long-term solution for the East Coast Greenway and that an interim route could be established using both existing local trails and on-street connectors.

On our third and final day we were joined by a larger contingent of riders, including Greg Hinchcliffe; Bob Patten; Jennifer Toole, owner of Toole Design Group, a Maryland-based bicycle, pedestrian, and greenway planning firm; and Martin Zimmerman, a writer for *Urban Land Institute* magazine. Our route of travel followed a new creekside trail just south of our hotel and traversed College Park, past the University of Maryland campus, to the Maryland-Washington, DC, jurisdictional boundary. Once we entered Washington, DC, our off-road options ended, and we wound through a network of neighborhood and city streets to Union Station. This was the most challenging aspect of our ride as traffic was heavy and bicycle accommodation was limited. We were able to see the first completed phase of the Metropolitan Branch Trail, built as part of a new elevated transit stop along the Washington, DC, Metro Green Line, and new bike lanes that DC was installing as part of its expanded on-street system. Bob Patten expertly guided our entourage into Union Station, and we arrived on time for a noontime celebratory event.

We completed our ride at Union Station with a media event that was attended by local supporters of the East Coast Greenway Alliance. Representatives from the Washington Area Bicyclists, Rails-to-Trails Conservancy, and American Trails were on hand to greet the riders. A brief press event was arranged by Bob Patten and included participants from the Washington, DC, City Council and Transportation Department. Much of the press event centered on the completion of the Met Branch Trail, a critical component of the off-road East Coast Greenway experience through the District. We also celebrated the placement of the Greenway Mid-Point Marker, which was installed at a new bicycle hub that was under construction adjacent to Union Station.

The author rides on the Baltimore to Annapolis rail-trail, one leg of the 3,000-mile East Coast Greenway. Source: Author.

To a person, the ride exceeded our expectations. We were pleasantly surprised at how easy to use and bicycle-friendly the majority of the route was. The landscapes and history of the region were of great interest to all the riders, and the ride served as a catalyst for action, helping the Maryland and DC state committees rally around the immediate development of off-road greenway segments.

The ride illustrated several important facts about the East Coast Greenway. First, how individual greenway and trail segments join together to create the envisioned 3,000-mile travel route. Second, the importance that local volunteers and municipal governments play in planning, designing, building, and maintaining greenway segments. Third, the level of excitement and engagement that is shared among the local communities that are officially a part of the greenway. The three-day journey reinforced the ability of the greenway to address trail users' desires to participate in an active healthy lifestyle, satisfy a close-to-home destination vacation opportunity, and provide alternative transportation options for those days when you might want to leave the car at home and ride or walk to work, school, or other community destinations.

A Seven-Day Ride: Wilmington, Delaware, to New York City

With the first ride completed, Karen Votava and I decided to sponsor another ride and expand what became known as the Chairman's Ride. In 2007, Karen and I accepted, on behalf of the ECGA, an invitation from a sister organization located in the Czech Republic to participate in a bike ride across the Moravian and Bohemian regions to help raise funds for the Czech Environmental Partnership. In 2001, I had hosted members of the partnership and the European Greenway Association as they toured American greenways hoping to learn from our successes. After years of planning, design, and development, the Czech greenway network was ready to host a small group of ECGA riders with the hope of showcasing their progress. Karen formed the Close the Gaps Club as a way of recognizing the largest funders of the ECGA. Each member of the club was extended an invitation to join the 2007 Czech Greenway Chairman's Ride.

The tour through the western part of the Czech Republic extended from Vienna to Cesky Krumlov. It was a spectacular five-day ride, and Karen and I immediately envisioned new possibilities for the East Coast Greenway. Karen returned from that experience with enthusiasm and renewed focus for hosting a similar ride along a section of the East Coast Greenway. What impressed us the most was how the Czech greenways had sought out hotels, restaurants, and other services along the route of the greenway and certified these businesses as friendly to greenway users. That meant for example that at hotels, you had a secure place to store your bicycle overnight. Some hotels even provided wash racks. Restaurants would accept casually dressed patrons, something that is important when touring by bicycle. Finally, the route of travel was well signed, especially where we navigated through the center of cities and towns and in some of the more remote rural locations. While we had a guided tour, it would have been easy to complete the same tour without a guide.

After six months of planning the ECGA staff, led by Eric Weiss, the ECGA national trails coordinator, I was ready to host a third Chairman's Ride that would emphasize the need to close additional trail gaps. Eric and the staff cobbled together an incredible route of travel that would take us from downtown Wilmington, Delaware, to Manhattan, through Philadelphia, Trenton, and Newark. It was an ambitious ride, and there was plenty of skepticism that our organization (the ECGA) was biting off more than we could handle at the time.

A seven-day ride along the East Coast Greenway is one that should be completed by more experienced cyclists, as you are more likely to encounter challenges in riding through urban areas. Among the many riders that joined the Third Chairman's Ride were Jim and Debbie Sharpe, who have for many years been among the most enthusiastic supporters of the East Coast Greenway Alliance. Jim Sharpe is entrepreneur in residence at the Harvard Business School, Arthur Rock Center. His wife, Debbie Stein Sharpe, is a graduate of MIT, with a degree from the Harvard Business School. They are successful in business and devoted parents to their three children. Jim and Debbie have traveled the world by bicycle, were members of our merry band of travelers to the Czech Republic, and were ready to take on the challenges of a similar adventure along the East Coast Greenway.

Our route of travel took us from the Christina Riverwalk in downtown Wilmington, along the Schuylkill River Greenway in Philadelphia, to the Delaware and Raritan Canal outside Trenton, through Princeton and Newark, to Jersey City and eventually around the island of Manhattan. The ride began on September 29 and concluded on October 5, 2008. Fortunately, the weather cooperated throughout the entire seven-day ride, which made each day enjoyable for bicycle touring. The total length of the trip was 225 miles, but due to several gaps in the off-road trail, we did not ride our bikes the entire way. We averaged about 25 miles of riding per day. Our seven-day trip up the east coast was remarkable for several reasons:

Heritage Landscapes

One of the most appealing aspects of traveling the route of the East Coast Greenway is that you can visit a variety of historic landscapes and industrial complexes that helped to build and shape America. Our ride began with a tour of the historic Brandywine River Valley in Wilmington, where we visited the DuPont Nemours "Homestead" museum, featuring the original gunpowder factories and machine shops. Farther north, just south of Philadelphia, we toured Fort Mifflin on the Delaware River, first established in 1771 and since the 1950s an important U.S. military post. In Trenton, New Jersey, we toured the Old Barracks Museum, established in 1758, which was celebrating its 250-year anniversary during our visit. Using the Delaware and Raritan Canal as our path of travel, we later that same day were treated to a vivid description and interpretation of George Washington's first victory in the Revolutionary War as we sprawled on the grass of Battlefield State Park outside Princeton, New Jersey. We rode past the famed Blue Bell Inn, established in Philadelphia in 1766, where it is rumored President George Washington once slept. We also experienced profound sadness as we visited Ground Zero in New York City, where the memorial to the victims of September 11, 2001, was under construction. The tour of significant American landscapes had a profound effect on each of the riders. This was made all the more meaningful by the sequence and intimate nature of our travel, as cyclists experiencing the landscape settings that were so different from each other.

Green Infrastructure

When you travel by bicycle over such a long distance, and across varied landscapes, you are much better connected by the experience of place than if you performed the same trip by car, bus, or train. The green infrastructure, in the form of parks, greenways, forests, wetlands, rivers, open meadows, and other natural landscapes along our ride, was a significant highlight of the entire trip. From parks planned and designed by Frederick Law Olmsted in Wilmington's Brandywine River Valley to modern parks along the west side of Manhattan, we experienced the beauty and appreciated the value that these landscapes contribute to urban communities. These greenspaces were typically full of people enjoying the sunshine, blue skies, and fresh air. We toured the estate grounds of the Cauffiel and Bellevue mansions in northern Delaware, now opened as components of the public parks system. The 1,000-acre John Heinz National Wildlife Refuge at Tinicum, Pennsylvania, transported us from urban life to a wilderness setting as we navigated the banks of Darby Creek, while in the distance we could see the skyline of center city Philadelphia. We toured famed natural scientist John Bartram's residence and garden just south of downtown Philadelphia. Bartram (1699–1777) was American botanist to the King of England. In New York City, we were all amazed by the Hudson River and East River transformation from industrial shipping port to an interconnected linear park system that rings the island of Manhattan.

Urban Culture

It was also fun to experience the transformation of center city and urban life in Wilmington, Philadelphia, Trenton, Newark, and New York City. The East Coast Greenway was a contributor to this transformation. These cities were undergoing substantial changes in urban form and cityscape so that they could attract new residents and reinforce the desired live, work, and play lifestyle. As a bicyclist traveling at modest speeds of fifteen to twenty miles per hour, we were afforded the opportunity to comprehend the transformation block-by-block as we wove our way through the changing urban fabric of each city. We noticed that the New York City shoreline was clean and bustling with thousands of residents and international tourists. Wilmington's Christina Riverwalk was buzzing with activity and dining and entertainment options. Philadelphia was investing heavily in the revitalization of the Schuylkill River waterfront. We witnessed firsthand what so many urbanists have been saying for years about American cities—that it is cool to live in city centers, there is plenty to do, and urban centers are replacing their aging industrial complexes with exciting multipurpose parks and greenways that improve the quality of life for area residents.

Since we completed the Chairman's Ride in 2008, many new off-road segments of the East Coast Greenway have been added between Wilmington, Delaware, and New York City. Most significantly, the City of Philadelphia was awarded two separate U.S. Department of Transportation, Federal Highway Administration TIGER grants (see chapter 8, "Callin' the Hogs," for more information about the TIGER program), which funded completion of the Schuylkill River Greenway Trail. New off-road greenway segments have been added to the route of travel in New York City, across the state of New Jersey, north and south of the city of Philadelphia, and in the state of Delaware. During the past decade, it

The Schuylkill River Greenway in downtown Philadelphia is representative of urban trails that comprise the East Coast Greenway. Photo by author.

has been a mile here and a mile there. That is how the East Coast Greenway satisfies the vision of an off-road trail, one mile at a time, knitting together a lot of local trails.

A One-Day Ride: The Cross Triangle Greenway

I imagine that most of the 11 million users of the East Coast Greenway experience the trail through hourly or daily rides. I participated in a half-day ride on a spectacular October day in 2009 to draw attention to the Cross Triangle Greenway. Alliance executive director Dennis Markatos-Soriano organized a forty-mile half-day ride on the greenway, from the North Carolina Museum of Art in Raleigh to the American Tobacco campus in downtown Durham. Approximately fifty people participated in the ride.

The Cross Triangle Greenway mirrors the East Coast Greenway itself, as a collection of local trails that have been linked together to provide a continuous route of travel. The trail comprises the Walnut Creek Greenway in Raleigh, North Carolina Museum of Art Trail,

Reedy Creek Greenway in Raleigh, Umstead State Park multiuse trail, Black Creek Greenway in Cary, White Oak Creek Greenway in Cary, and the American Tobacco Trail, which extends through Wake, Chatham, and Durham Counties, and has its northern terminus in downtown Durham, at the Durham Bulls Baseball Stadium and the American Tobacco Campus.

We departed around 9:00 a.m. from the Museum of Art, which was undergoing renovations at the time, and immediately made our way west to Umstead State Park, utilizing a portion of Raleigh's five-mile long Reedy Creek Greenway. We passed through a portion of North Carolina State University's Schenck educational forest and entered the state park, where we transitioned from the paved portion of Raleigh's Reedy Creek Greenway to the stone dust–paved multiuse Reedy Creek trail.

Crossing over Interstate 40 proved tricky as there was no direct connection with the town of Cary's expansive greenway system at the time. We used a very busy North Harrison Avenue, riding past the international headquarters of SAS Institute, one of North Carolina's largest employers. From North Harrison Avenue, we connected with Weston Parkway and entered the Black Creek Greenway just south of Lake Crabtree, a regional flood control project. My firm had designed a portion of the Black Creek Greenway, and it had been years since I had bicycled that section of the greenway. All of the riders stopped for a short break in North Cary Park where the town staff had set up a refreshment stand, and several members of the town's park staff were on hand to make a presentation about Cary's greenway system and their plans for closing gaps in the Cross Triangle Greenway. It was educational for all of the riders, some of whom had never even heard of the Cross Triangle Greenway or the East Coast Greenway.

Continuing on the Black Creek Greenway for several miles, the off-road trail terminated at NW Maynard Road, and we had a short ride along bike lanes over to Cary's premiere park, the Fred G. Bond Metro Park, which has an extensive trail system. After leaving Bond Park, we picked up another off-road greenway that was also under construction, the White Oak Greenway. This developing trail is routed through Cary's fastest-growing suburban neighborhoods, extends under Interstate 540, also known as the Triangle Expressway, and eventually will provide a direct connection to the American Tobacco Trail in southern Wake County.

It was getting close to lunchtime as our group arrived at the end of the White Oak Greenway. We had to take a short detour along rural roads to the American Tobacco Trail Wimberly Road Trail Head. We were once again treated to a refreshment stop. Wake County Parks and Recreation staff and ECGA staff provided a brief explanation of the American Tobacco Trail as riders continued to flow in and out of the trailhead.

The American Tobacco Trail was North Carolina's first long-distance rails-to-trails conversion. My company prepared the master plan for the trail in 1992, and it took about twenty-five years for the project to be completed. The trail is twenty-two miles long and follows a corridor that was once known as the New Hope Valley Railroad and later the Durham and South Carolina Railroad. This railroad corridor began as a short line extending from downtown Durham into the New Hope Valley watershed, where large stands of old growth forest were harvested for lumber. Decades after its original founding, the rail line was extended south to connect with the Norfolk-Southern's east-west North Carolina railroad so that tobacco could be shipped from eastern North Carolina farms into the American Tobacco processing plants in downtown Durham.

The line was relocated by the U.S. Army Corps of Engineers in 1969 to accommodate future development of the B. Everette Jordan Dam and Reservoir. By the mid-1970s the relocated line was abandoned after Southern Railway and Norfolk-Southern merged, making the line redundant. Formal abandonment began in 1979, and the corridor remained a long dirt road for more than a decade when my firm was approached by the Triangle Rails-to-Trails Conservancy to begin work on a greenway master plan.

Traveling the route from the south, I quickly realized the slight strain of riding the rail corridor. It is a slow ascent along the entire twenty-two miles as you climb in elevation from Wake County through Chatham County and into downtown Durham. We made a brief stop at the Streets of Southpoint Mall. The REI Durham store located across from the Mall was hosting the riders with another pit stop. REI technicians were on hand to help riders with their bikes, and additional refreshments were available inside.

Our group pushed on, and we had to make use of the very busy Fayetteville Road crossing of Interstate 40, as there was no dedicated greenway bridge at the time crossing over the interstate (since that ride, the City of Durham constructed a bicycle and pedestrian bridge that spans the interstate). North of the interstate, we quickly rejoined the American Tobacco Trail for the remaining 6.5 miles into downtown.

The ECGA rolled out the red carpet for the riders, and we were greeted by David Price, U.S. Congressman for North Carolina's Fourth District, and North Carolina Department of Transportation Secretary, Gene Conti (a very good friend and colleague). After a few short speeches, the riders were encouraged to continue their journey north into downtown Durham. The ECGA staff had arranged a bus and bike transfer back to the North Carolina Museum of Art.

The Cross Triangle Greenway ride showcased the daily utility of the East Coast Greenway for commuting, health, and wellness within city and metro areas. In North Carolina, the trails that comprise the Cross Triangle Greenway are already some of the most heavily traveled in the state. Branding the route of travel as an element of the East Coast Greenway has become a source of pride for Raleigh, Cary, and Durham, as all of these communities expand their greenway networks.

The Future of the East Coast Greenway

The East Coast Greenway is an American success story. From humble beginnings envisioned by a handful of people, the greenway today is embraced and enjoyed by millions. Unlike most other long-distance trails in the United States (especially those that are 1,000 miles and longer), the East Coast Greenway is located within close proximity of where millions live, work, and recreate. As such, it offers daily benefits that other long-distance trails cannot. The greenway supports an active lifestyle, which benefits public health. As green infrastructure, the corridor conserves land, mitigates air pollution, intercepts stormwater runoff, and provides habitat for urban wildlife. The greenway is also part of a resilient and sustainable landscape that adds long-term value to its host communities. The return on investment in trail development pays dividends every day. Importantly, the greenway is a democratic landscape, open for use by all people regardless of race, gender, or socioeconomic condition.

In 2017, the East Coast Greenway celebrated its twenty-fifth anniversary. The greenway has grown and matured over time into one of the most admired green infrastructure projects in the United States. As of 2018, the greenway attracted more than $1 billion in public and private investment. The number of people involved in greenway planning, design, development, operation, management, and event programming continues to grow exponentially, both at the ECGA and in the myriad partner communities along the route.

Without a significant infusion of public sector funding, the greenway is most likely decades from completion. In 2018, the ECGA coordinated with its trail development partners and submitted applications to the U.S. Department of Transportation for more than $100 million in funds to complete 200 miles of new off-road greenway in seven states. The ECGA, working hand-in-hand with its partners, continues to add new off-road trail miles every year. However, significant challenges remain in completing the route of travel. In the New England region, from Maine to Connecticut, the easiest sections of the greenway have been completed. The challenging sections of the trail must overcome lack of available right-of-way, river and highway crossings, and routing solutions through dense urban development. The Mid-Atlantic region, from New York City to Washington, DC, faces similar challenges to the New England region. The South Atlantic region, from Washington, DC, to Georgia, faces completely different challenges, in the form of new long-distance off-road trails stretching across suburban and rural landscapes. Hundreds of miles of right-of-way must be secured and long-distance off-road trails constructed. The Southeast region, Georgia and Florida, is challenged to complete final route and alignment through both rural landscapes in Georgia and urban areas in Florida.

The Alliance estimates that $2 billion in funding is needed to finish the remaining 2,000 miles of off-road greenway, an average cost of $1 million per mile. That sounds daunting and perhaps unattainable until you put into perspective the billions in economic return that would be realized from its completion. The greenway represents one of the most fiscally sound investments that the United States can and should make in green infrastructure development.

The Experience Economy

The greenway is representative of an evolving global economy. In 1999, economists Joseph Pine and James Gilmore theorized that future economic growth would occur in the value of experiences rather than limited to goods and services produced. They believe that in order to sustain future economic growth, what they termed the "Experience Economy" would be an important part of future business transactions, with a focus on which businesses orchestrate memorable events for customers.[4] The companies leading the charge in this new economy are now some of the largest corporations in the world: Apple, Google, Facebook, and the like. The acceptance and development of the East Coast Greenway reflects the value today's consumers place on the emerging Experience Economy.

The benefits of the Experience Economy are not limited to the products, services, and experiences provided by corporations; all types of business and personal decisions are influenced, including the desire to purchase homes, cars, and other items such as high-end bicycles, or the allocation of household dollars to experiences versus debt—for example, the desire of millennials to celebrate a night on the town with friends versus their desire to pay a mortgage. The East Coast Greenway supports the Experience Economy.

North-South Travel and Tourism Corridor

The Experience Economy also provides financial support for the future of the East Coast Greenway as a sought-after national and international destination landscape for short-distance and long-distance bicycle touring. The one-day, three-day, and seven-day rides that I described demonstrate the potential of the greenway in this regard. This is the long-term view of the greenway, as more and more states complete their sections and longer, interconnected routes are established. The ECGA itself might evolve into a nonprofit organization more focused on coordinating programming and travel itineraries for thousands of greenway riders.

The greenway has already achieved the status as one of the most significant north-south urban spine trails in America. Equally significant is the fact that the greenway is also quickly becoming the signature spine trail for the 450 communities through which it travels. Local communities are quick to state that they have a section of the East Coast Greenway extending throughout their jurisdiction.

The greenway has the ability to transform rural communities, providing much-needed business incubation, job creation, and sustainable economic growth for areas of the East Coast that definitely need a financial windfall. In urban centers, the greenway will continue to be an important element of revitalized landscapes that celebrate the cultural diversity that has always been a hallmark of the United States.

One day hopefully in the not so distant future, a young boy (with permission from his mother) will fulfill his dream of riding a bike safely along the East Coast Greenway from south Philadelphia to New York City and will accomplish the ride with family and friends, either as part of a once in a lifetime excursion or as an annual family tradition. Therein lies the vision and the promise of the East Coast Greenway.

11

A National Greenway System

Envisioning a Coast-to-Coast Greenway System

The Indianapolis Cultural Trail: a legacy of Gene and Marilyn Glick. Source: Indianapolis Cultural Trail, Inc.

THE STORIES IN THIS BOOK help illustrate how greenways evolved from their nineteenth-century origins as linear connectors of city parks to twenty-first-century strategic essential green infrastructure planned, designed, and constructed to conserve irreplaceable greenspace, improve mobility, promote a healthy lifestyle, and protect public health, safety, and well-being. Greenways come in many shapes and sizes, are found in a variety of environmental settings, and are a landscape typology used to address and resolve a wide variety of environmental and societal issues— as varied as the landscapes and the people that develop them!

The Anne Springs Close Greenway is testament to the impact that a passionate, forward-thinking individual can have on land conservation, community development, and regional identity. This greenway illustrates how philanthropy provides opportunities to give back and pay forward the gift of greenspace conservation. The Red River Greenway demonstrates how communities can recover from natural disasters to become more resilient to repetitive flooding. This greenway has become the most important public asset of the Greater Grand Forks region, helping to absorb flooding, support economic development, and improve the quality of life for residents. The Swift Creek Recycled Greenway transforms the way we think about our waste stream and recyclables, as well as the life cycle of raw materials and single-use products. The Grand Canyon Greenway resolves visitor conflicts by providing a greenway network that offers safe, efficient, and accessible paths of travel for all people regardless of their abilities and mobility, while enhancing the experience of place for one of the most visited landscapes on the planet.

In two major metro areas, Las Vegas, Nevada, and Miami, Florida, greenways transformed undervalued landscapes into community assets. Charleston County, South Carolina, discovered ways to leverage local funding to conserve, protect, and link together its cherished Lowcountry landscape. In Northwest Arkansas, the Walton Family Foundation developed greenways to improve health and wellness, support diversified economic development, and create a robust bicycling and walking transportation network.

The nation of Belarus envisioned using greenways to diversify its agriculture-based economy. The same can be accomplished across the American Heartland. Americans should more readily embrace the benefits that greenways can bestow on rural communities. The East Coast Greenway demonstrates the power and appeal of a grand vision to link hundreds of cities and towns together with long-distance bicycle and pedestrian paths to improve travel and tourism while conserving valued landscapes that celebrate American history and culture.

These stories represent core planning and design strategies that provide the basis for how future greenways will continue to develop in American communities for many years to come. What is the future of greenways in the evolving American landscape? What societal imperatives, opportunities, and challenges will greenways help to address and resolve? How do we leverage the greenway accomplishments of the past to build more resilient and sustainable communities in the future?

In the previous ten chapters, I shared my stories about ten exemplary greenway projects. In this chapter, I will describe a bold vision for implementing a national greenway system. Imagine the possibilities. The reality is, America, over the decades, has been building up to a national interconnected greenway network, one corridor, one trail, and one project at a time. The nation as a whole, however, has not paid much attention to the reality of a national greenway system. To better understand the future of greenways, let's step back in time to consider a proposal introduced more than 100 years ago by landscape architect Warren Manning. In 1919, Manning completed a National Plan envisioning a future for the United States that included a network of greenways linking together national and state parks and conservation areas. He proposed connecting corridors spanning the nation called "recreation ways," the purpose and intent of which are consistent with how we define modern-day greenways. How close are we to realizing Manning's vision for an interconnected national greenway network? Is Manning's vision a viable platform on which to build a national interconnected greenway network for the United States? Let's unpack the historical context and framework of a national network of trails and greenways and define the steps to be taken to realize full build-out.

One of the critical issues that a national network of greenways must address are the impacts and consequences of global climate change on natural ecosystems and human society. America's ecosystems and microclimates are changing rapidly due to warming of the planet. We continue to experience more frequent extreme weather events, punctuated by increased rainfall and severe flooding in some parts of the nation and simultaneously devastating severe drought and wildfires in other parts. Rapid warming and extreme weather impact the health, safety, and well-being of our communities, particularly in shoreline areas, while simultaneously impacting native plant and animal species. We need an interconnected green infrastructure system to address the needs and concerns of humans, plants, and animals. We must consider the benefits associated with a national framework of interconnected greenways as "gene-ways." What does this framework look like, and what steps must be taken to build a cross-country network of gene-ways?

From the perspective of American culture, a national system of trails and greenways acknowledges the richness and opportunities that celebrate America's cultural diversity, ethnic origins, and racial composition. To accomplish this, greenway planning and design must become more inclusive of varying viewpoints and cultural identities, throughout all phases of development. If we recognize that the traditional and historic vision of greenways, and their value to American society, is not the only relevant perspective, then we must envision and implement methods of public engagement that empower and enable traditionally underserved communities to benefit from greenway development. How might greenways better serve the needs of diverse populations?

Historic Context for a National Greenway Strategy

Make no little plans; they have no magic to stir men's blood and probably themselves will not be realized. Make big plans: aim high in hope and work, remembering that a noble, logical diagram, once recorded, will never die.

DANIEL BURNHAM, 1907

Warren Manning's "National Plan"

Almost in direct response to Burnham, landscape architect Warren Manning in 1919 authored "A National Plan." This 565-page tome defines a visionary and comprehensive development strategy for the continental United States. Manning's "Plan" addresses a myriad of issues and strategies including:

A description of America's bountiful forests, grasslands, mineral, land, water, and power resources,

A prediction of population growth and where it would be distributed (in 1916 American population exceeded 100 million),

Recognition of the importance of scenic, natural, and human-made resources, including a national "recreation system,"

A visionary national transportation network comprising rail, interstate roadways, and "aeroplanes" (new mode of travel at the time),

Plans for how major cities should grow, and

Detailed strategies and plans for the forty-eight continental states.

Manning concluded: "If we want our country to stand first among all nations then very many more of our citizens must gain a better understanding of this United States as a whole. We must lay down and execute a national plan."[1]

The context of American life during the time that Manning authored his "National Plan" is important to understand. In 1917, President Woodrow Wilson sought authorization from Congress to commit American troops and resources in support of European allies who were fighting Germany in the Great War. Transportation in America was rapidly evolving, transitioning from dependence on horse, streetcar, and rail as primary means of mobility to the automobile and a system of national highways. A total of 1.6 million cars were manufactured and sold in 1916, changing the way Americans traveled throughout cities, towns, and countryside. The City of Dallas, Texas, opened a municipal airport, Love Field, to service the takeoffs and landings of the "aeroplane," a Wright Brothers invention that sparked interest and enthusiasm for air travel.

In short, this was a time of major changes throughout America. For Manning, the convergence of interests and activities represented an opportunity to author a visionary plan for America that combined the nation's emerging status as a world leader with its bountiful resources and thirst for invention and innovation. Manning's vision for a national network of greenways spanned the continent, linking an emerging system of national parks, forest preserves, and state-owned lands. It was to be a combined system based on an ecological framework as well as travel and tourism nodes and corridors. Manning further envisioned that "the Recreation Way will use existing roads [a national network of roads was a developing concept at the time] where they follow lines of greatest beauty or new roads will ultimately be established to make connections and to make accessible new scenic regions."[2]

Manning was not a novice when it came to understanding interrelationships between landscape and human life. As a protégé of Frederick Law Olmsted Sr., whom he apprenticed under from January 1888 to December 1895, Manning assisted Olmsted with some of his most important commissions, including the Stanford University Master Plan in Palo

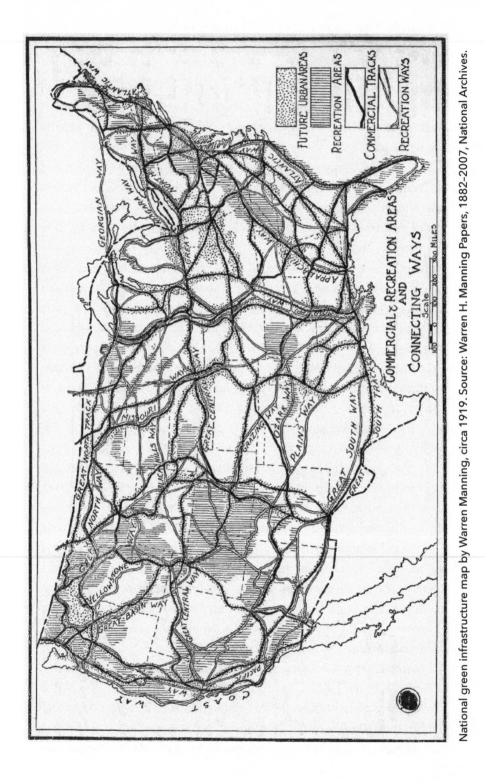

National green infrastructure map by Warren Manning, circa 1919. Source: Warren H. Manning Papers, 1882-2007, National Archives.

Alto, California; the 1893 Columbia Exposition in Chicago, Illinois; and Biltmore, George Vanderbilt's Asheville, North Carolina, estate. Manning left the firm of Olmsted and Sons in 1896 to open his own practice, where he became nationally known and highly regarded for his extensive knowledge of horticulture and landscape planting design. While Manning's work focused on the estates and gardens of wealthy clients in the northeastern United States, he never lost his interest or appetite for the larger landscape context.

Manning was a contemporary of Olmsted's son, Frederick Law Olmsted Jr., who worked closely with California Congressman William Kent to author key elements of the 1916 Organic Act, which led to the establishment of the National Park System. Regarding a national greenway network, Manning concluded: "If we are given favorable opportunity for such out-door recreation for all our people we must extend our recreation spots and patches, and those indicated on the Recreation map in so much of our territory, into a system so devised as to enable all the people to find a place near their homes where well organized, low cost transportation, camping and food supply facilities, will enable them to take frequent advantage of it."[3]

Benton MacKaye's Appalachian Trail

Shortly after Manning completed the "National Plan" manuscript, Benton MacKaye, a forester, conservationist, and regional planner, envisioned the Appalachian Trail, a single corridor of conserved land that would extend 2,100 miles from Maine to Georgia along the Appalachian Mountain range. MacKaye proposed the concept of land preservation for recreation and conservation purposes. He, like Manning, understood that there was the need to balance human activity with the beauty and resources of nature. Both visionaries were concerned with the rapid industrialization and population growth in the eastern United States and the resultant impacts on natural resources. MacKaye's proposal for the Appalachian Trail was published in the October 1921 issue of the *Journal of the American Institute of Architects,* titled "An Appalachian Trail: A Project in Regional Planning." He envisioned a wilderness hiking trail that would link together large patches of conserved "wilderness communities," where the average American could go to "renew body, mind and spirit." In many ways, MacKaye's vision represented one corridor type within Manning's national greenway network. Over time, the trail portion of MacKaye's vision became a reality; the wilderness retreats did not. To this day the Appalachian Trail remains one of America's great achievements of conservation and trail development. It is widely regarded as the nation's first long-distance greenway that establishes a model framework for implementing other long-distance greenway projects.

Manning's visionary "National Plan" remains an unpublished manuscript. It is included as a portion of his professional archive at the Iowa State University Library and the National Archives in Washington, DC. Had Manning published his plan, it might have been completely lost in the overwhelming events of the day. The aftermath of World War I across Europe consumed the attention and resources of America. A decade later the world was rocked again by the 1929 Stock Market Crash, which triggered a global economic collapse resulting in the Great Depression. Just as economic recovery began to take hold in 1939, Nazi Germany invaded Poland, launching World War II. In September 1945, the second World War officially ended in Tokyo Bay on the deck of the American battleship USS *Missouri*. Two years later global tensions again heated up as the Cold War ensued among former allies, the United States and Britain at odds with the Soviet Union. Communism spread across the Korean Peninsula in early 1950, resulting in the Korean War and creating the lasting divisions of North and South Korea. In 1955, a second Indochina War began in Vietnam that consumed American interests through three presidential administrations, eventually ending with the Paris Peace Accord in January 1973.

The National Trails System Act

Throughout this fifty-five-year period, few Americans focused on or even knew of Manning's vision for an interconnected national trails and greenway system. However, beginning in 1964, under President Lyndon Johnson's leadership and vision for a Great Society, the United States began to focus on domestic policies and programs geared toward improving the quality of life for every American. In a February 1965 speech to Congress focused on "Conservation and Preservation of Natural Beauty," Johnson prescribed a new national strategy for trails and greenways. He observed:

"As with so much of our quest for beauty and quality, each community has opportunities for action. We can and should have an abundance of trails for walking, cycling, and horseback riding in and close to our cities. In the back country we need to copy the great Appalachian Trail in all parts of America, and to make full use of rights-of-way and other public paths."

That same year, in September 1965, Wisconsin Senator Gaylord Nelson introduced legislation in the 89th Congress to establish a national hiking trail system, stating, "In order to provide for the ever-increasing outdoor recreation needs of an expanding population and in order to promote the preservation of public access to travel within and enjoyment and appreciation of the open-air, outdoor areas and historic resources of the Nation, trails should be established (i) primarily, near the urban areas of the Nation, and (ii) secondarily, within scenic areas and along historic travel routes of the Nation which are often more remotely located."

Stewart Udall, secretary of the interior under Presidents Kennedy and Johnson, was an important visionary and one of the most conservation-minded public officials in American history. In 1963, he authored *The Quiet Crisis*, which warned of the dangers of pollution, exploitation of natural resources, and rapidly disappearing open space. In December 1966, he released *Trails for America*, one of the most significant greenway proposals since Manning's 1919 manuscript. As interior secretary, he was able to expand federal public lands and gain passage of significant environmental legislation, including the Clean Air Act, Clean Water Act, Wilderness Act of 1964, Endangered Species Act of 1966, Land and Water Conservation Fund Act of 1965, and National Trails System Act of 1968.

The 1968 National Trails System Act provided the legislative foundation for a national interconnected greenway network that defined three types of nationally significant trails in the system:

> National Scenic Trails: trails of more than 100 miles in length that provide for outdoor recreation and "for the conservation and enjoyment of the nationally significant scenic, historic, natural, or cultural qualities of the areas through which such trails may pass." National Scenic Trails are land-based, necessarily excluding water-based travel routes. These trails are designated and authorized by an Act of Congress.

National Recreation Trails: trails less than 100 miles in length that
follow historic trails or routes of travel as closely as possible. The
purpose of these trails is "the identification and protection of the
historic route and its historic remnants and artifacts for public
use and enjoyment." National Historic Trails, unlike National
Scenic Trails, may include water-based routes. Examples include
the Red River Greenway in Grand Forks, North Dakota, and the
American Tobacco Trail in North Carolina, two trails that I had
the honor to plan and design.

Connecting and Side Trails: provide opportunities for outdoor recreation primarily in and around urban areas and have no mini-
mal length requirement. These trails may be designated by either the secretary of the interior or the secretary of agriculture
rather than by an act of Congress. These trails may exist entirely
on state, local, and private property as well as on federal lands.[4]

In 1978, President Jimmy Carter added National Historic Trails as a fourth
type of trails to the National Trails System Act, which recognized original trails or routes of travel of national historic and cultural significance,
including past routes of exploration, migration, and military action. The
National Trails System Act also established the game-changing concept
of railbanking, which provides America with the legal framework for
conserving abandoned railroad rights-of-way and led to the establishment of an incredible national network of rail-trails.

The President's Commission on Americans Outdoors

In 1985, President Ronald Reagan appointed the President's Commission
on Americans Outdoors (PCAO). This was an important turning point
in American history for greenway development, which provided critical foundational support for a national system of interconnected greenways. This commission was charged with reviewing public and private
outdoor recreation opportunities, policies, and programs and making
recommendations to ensure the future availability of outdoor recreation
for the American people.[5]

The appointed members of the commission included Lamar Alexander
(chairman), governor of Tennessee; Gilbert Grosvenor (vice chairman),

President Lyndon B. Johnson and the National Trails System Map, circa 1968. Source: U.S. Department of the Interior, National Park Service.

president of the National Geographic Society in Washington, DC; Frank Bogert, mayor of Palm Springs, California; Sheldon Coleman, chairman of the Coleman Co., Inc., in Wichita, Kansas; Derrick Crandall president and chief executive officer of the American Recreation Coalition in Washington, DC; J. Bennett Johnston, a United States senator from Louisiana; Charles Jordan, director of parks and recreation for the City of Austin, Texas; Wilbur LaPage, director of the division of parks and recreation for the State of New Hampshire; Rex Maughan, president and chief executive officer of Forever Living Products, Inc., in Phoenix, Arizona; Patrick Noonan, president of the Conservation Fund in Arlington, Virginia; Stuart Northrop, chairman of the board of Huffy Corp. in Dayton, Ohio; Morris Udall, a member of the U.S. House of Representatives from Arizona; Barbara Vucanovich, a member of the U.S. House of Representatives from Nevada; and Malcolm Wallop, a United States senator from Wyoming.

The PCAO's mission was the continuation of important work begun in 1958 by the Outdoor Recreation Resources Review Commission

(ORRRC) under President Dwight Eisenhower. The ORRRC was then charged with completing an inventory and evaluation of outdoor recreation resources using information obtained from federal, state, regional, and local governments. The goal of ORRRC was to assess how a new generation of post–World War II Americans accessed and used America's vast natural resources and recommend strategies for improving outdoor access for all Americans. Laurance Rockefeller, one of the nation's most important conservationists (named "Mr. Conservation" by President George H. W. Bush when he presented Rockefeller the Congressional Gold Medal for his lifetime of service), was chair of the ORRRC and is credited with defining a "conservation ethic in America." In January 1962, under Rockefeller's leadership, the ORRRC submitted a final report to President Kennedy titled "Outdoor Recreation for America." The findings of ORRRC formed the basis for how federal, state, and local governments understood the value of America's outdoors that: (1) walking for pleasure or exercise and picnicking were the most popular outdoor activities; (2) more outdoor opportunities were most needed near metropolitan areas; (3) considerable land was available for outdoor recreation, but it did not effectively meet existing needs; (4) outdoor recreation was compatible with other resource uses, such as management for wildlife and watersheds; (5) water was a focal point of outdoor recreation; and (6) outdoor recreation would bring about significant economic benefits.[6]

Almost twenty years had elapsed since the ORRRC released its findings by the time Ronald Reagan became president in 1980. Prior to his becoming president, a group of conservation leaders held meetings with members of the United States Senate with the hope of renewing the ORRRC and completing an updated evaluation of its findings and conclusions. This group included Laurance Rockefeller, Patrick Noonan, president of the Conservation Resources Group, Dr. Emery Castle, president of the Resources of the Future, William Reilly, president of the Conservation Foundation, Senator Malcolm Wallop of Wyoming, Henry Diamond, a friend of Laurance Rockefeller and noted environmental lawyer, Sheldon Coleman, president of the Coleman Corporation, and William Penn Mott, president of the California State Parks Foundation. As a result of more than a year of meetings, the group of conservation leaders concluded that two decades (1960–1980) since the formation of ORRRC had been a period of great social and economic change in how and where

Americans lived, worked, and played, resulting in profound effects on outdoor recreation. By 1980, outdoor recreation became more important in American life than it was in 1962. Local, regional, and state government agencies did less for outdoor recreation than what was required to meet America's need; the private sector, however, provided more outdoor recreation than it did in the prior eighteen years and could do even more in partnership with the government.

In August 1985, President Reagan was persuaded by Laurance Rockefeller, Senator Wallop, and others to appoint members to the PCAO. The commission operated for sixteen months from August 1985 to January 1987, conducting eighteen public hearings, along with eleven strategic planning sessions. The PCAO generated a number of recommendations, the most significant of which jump-started the American greenway movement. The commission's exact recommendation was as follows: "The Commission envisions a new system of 'greenways' along scenic highways, waterways and trails, linking communities and cities and the expansion of the federal estate."[7]

The PCAO believed that American communities should establish greenways—corridors of private and public recreation lands and waters—to provide people with access to open space close to where they live and to link rural and urban spaces in the American landscape. This was the first time a federally appointed group recognized the concept of greenways as a land use objective. Interestingly, the vision of the PCAO was very similar to the recommendations featured in Manning's "National Plan," authored sixty-five years earlier.

The "expansion of the federal estate" drew scrutiny and criticism, delayed the acceptance and release of the commission's report, and resulted in lawsuits and accusations leveled at the commission of improper procedure. The overarching theme of Ronald Reagan's two terms as president was to reduce the role of the federal government in the lives of Americans. The PCAO put the president in an awkward position by recommending the opposite course of action. PCAO chairman Lamar Alexander was convinced that the results of the commission's work called for a "prairie fire of action" to better accommodate America's need for and access to outdoor resources. The PCAO, therefore, viewed greenways as igniting that "prairie fire of interest" and, as an appeal to local interests, sought to mobilize citizens across the nation. Greenways, according to the PCAO, were landscapes that any American could conserve and steward in their backyard, neighborhood, community, and region. Commission vice chairman Gil Grosvenor, president of the National Geographic Society, was so taken with the concept of greenways he directed staff writer Noel Grove and photographer Phil Schermiester to prepare a feature article on the subject for the *National Geographic* magazine (published in June 1990). According to *National Geographic*, the greenways article received one of the highest levels of response in the history of the magazine. PCAO Commissioner Patrick Noonan did not need a whole lot of convincing that greenways were an important landscape typology for America. Shortly after the dust settled on the release of the PCAO report, he commissioned Charles Little to author *Greenways for America*. Upon completing the book, Little further recommended to Noonan the need for a "soup to nuts" book that would define how greenways were planned, designed, and constructed. Noonan subsequently hired Bob Searns and me to author *Greenways: A Guide to Planning, Design and Development*. At this time, David Burwell and Peter Harnik were in the process of launching the Washington, DC–based Rails-to-Trails Conservancy. Burwell had been working for the National Wildlife Federation in 1985 when he envisioned the opportunity to save abandoned railroad corridors and transform them into trails as a method of railbanking them for future transportation use. The collective sum of these actions thus launched the American greenway movement that gained significant momentum across the nation.

Intermodal Surface Transportation Efficiency Act, 1991

Prior to 1990, there was no dedicated funding to support greenway development. That changed in 1991 when President George H. W. Bush signed into law the Intermodal Surface Transportation Efficiency Act, known by its acronym as ISTEA ("ICE TEA"). This was the most important update to President Dwight Eisenhower's 1956 Surface Transportation Act, which resulted in the development of America's interstate highway system. ISTEA spawned a new era in and approach to transportation planning, design, and development, which focused on intermodal travel and transportation alternatives to automobiles, such as transit, bicycling, and walking.

The key leaders in the crafting of ISTEA included Congressmen Bud Shuster of Pennsylvania and Norman Mineta of California, and Senators Patrick Moynihan of New York, Harry Reid of Nevada, and Lloyd Bentsen of Texas. Idaho Senator Steve Symms proposed the establishment of the National Recreational Trails Trust Fund, which established dedicated revenues for developing motorized and nonmotorized trails from non-highway recreational fuel taxes. The provision authorizing the program was renamed the Symms National Recreational Trails Act of 1991; Senator Symms announced in August 1991 that he would retire at the end of the 102nd Congress. (I stood alongside Senator Symms and other national trail advocates as the Symms Act was introduced into Congress as an official element of ISTEA.)

As a result of ISTEA, federal transportation funding has been the largest single funding source fueling the growth and expansion of greenways across the nation. Prior to ISTEA, a total of $50 million was spent in support of bicycle-facility development. Since the passage of ISTEA (over a period of twenty-seven years), more than $10 billion has been spent to fund the development of greenways, trails, and various on-road bicycle and pedestrian projects!

The Modern National Greenway System

By the late 1990s and into the beginning of the twenty-first century, the acceptance of greenways coincided with important demographic changes in the United States. The nation was completing a transformation from one where most Americans lived in rural communities to one where the majority lives in urban communities. This gave rise to the importance of greenways as a landscape that could respond to a myriad of urban and suburban issues, such as conservation, green infrastructure, transportation, health and wellness, and economic development. This in turn established a foundation on which Americans could realistically complete a nationwide network of greenways. The resultant network has emerged in a manner that Warren Manning envisioned, with large tracts of federal and state lands linked by corridors of land and trails that spanned the nation. By combining the legislative efforts of the 1960s with landmark funding from the 1990s, a federally endorsed framework was established to support building the network in a uniform manner across the nation.

During this time frame, greenways became an identifiable and familiar landscape that Americans wanted to develop in their community. American citizens could state, with explicit understanding, that they wanted to build a greenway in which the fundamental ideals of conservation and trail development were embodied.

Between 1985 and 2018 American communities built and opened for public use an estimated 50,000 miles of new greenways and trails, the magnitude of which is roughly equivalent to the size of the American interstate highway system (which has always been thought of as one of the great public works undertakings in human history). This conservative estimate of accomplishment includes approximately 30,000 miles of converted abandoned railroad corridors, 5,000 miles of various state-sponsored greenways, 7,500 miles of regional greenways, and 7,500 miles of municipal greenways. If one adds to this total the 193,500 miles of trails already in place on federally owned lands and the 42,500 miles of trail that are located on state-owned lands, then America has the making of a national trails system that is approximately 280,000 miles in length![8] The basis for a very rough coast-to-coast, interconnected greenway system therefore exists. With concentrated efforts, gaps in the system can be evaluated, planned, constructed, and closed to create a completed national greenway network.

Next Steps in Completing the System

More than one hundred years has elapsed since Manning completed his "National Plan" that included the framework for a national greenway system. So now what? Where do we go from here? How can America build on the trails and greenways foundation of the past 100 years? Today, a new coalition of greenway advocates and leaders is emerging and poised to build on the accomplishments of the past, to envision and execute new strategies that will complete the national greenway network. The coalition, loosely organized as a public-private alliance of interests, represents local, regional, state, and federal governments, along with representatives from design and construction, business, philanthropic, and nonprofit sectors. While there is no single leadership structure for this alliance, organizations like American Trails, the Rails-to-Trails Conservancy, and the East Coast Greenway Alliance from the private and nonprofit sectors,

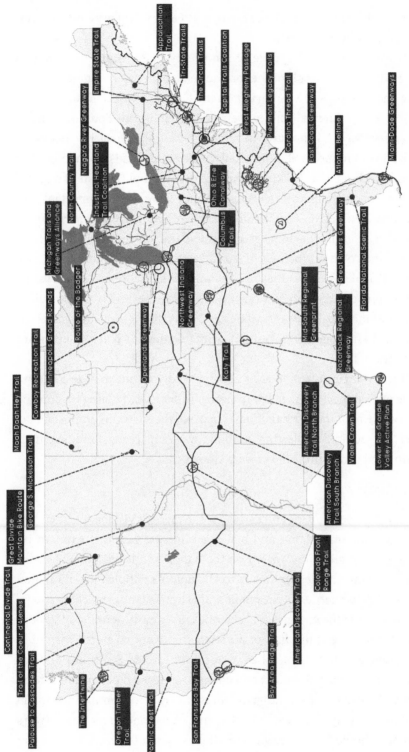

National Trails and Greenway Map. Source: Author and Pennsylvania Environmental Council, 2019.

along with the National Park Service and Federal Highway Administration, are coalition leaders.

During the second decade of the twenty-first century the coalition strengthened as a result of the formation and success of regional greenway and trail organizations, such as the Carolina Thread Trail in metro Charlotte, the Circuit Trails in metro Philadelphia, the Southeast Michigan Greenway Collaborative in metro Detroit, the Great Rivers Greenway District in metro St. Louis, and the Bay Area Trails Collaborative in metro San Francisco. Seeking to further the accomplishments of regional greenways, the Rails-to-Trails Conservancy launched an important programmatic initiative, TrailNation, that shares information and resources among a handful of regional greenway and trail organizations. The interests of the coalition focused in two areas: first is the buildout of metro and regional greenways that satisfy local demand for land conservation and improved access to outdoor resources; and second is the pursuit of long-distance regional, intrastate, and interstate greenways that connect and knit together, forming the interconnected national greenway network.

There remains much to accomplish in order to fully realize a nationwide interconnected greenway network. To achieve Manning's vision and to complement the work that has occurred, the following important actions are required: leadership, funding, system mapping, branding, marketing and promotion, and closing system gaps.

Leadership

During the 1960s, 1980s and early 1990s, America's visionary leaders understood the importance and beneficial impacts resulting from an interconnected greenway system on the nation and its economy. Needed are new American leaders ready to champion a national greenway network. They are needed in the public and private sectors, and at the federal level, to guide the energies and activities of a new generation of Americans toward the realization of a national network of greenways. Since the tragic events of September 11, 2001, the United States has directed substantial capital, attention, and energies on foreign policy and foreign wars. The twenty-first century requires America to commit investment to domestic gray and green infrastructure. A national greenway network can and should be a strategic and vital element of this effort, as it was under Presidents Eisenhower, Kennedy, and Johnson in the 1950s and 1960s and Presidents Reagan, Bush, and Clinton in the 1980s and 1990s. Federal leadership and investment need to match the commitment and leadership coming from the private sector.

Funding

From 1991 to 2009, transportation funding fueled the growth and development of community, regional, state, and federal greenway projects. America needs a dedicated funding source of equal value to complete a national greenway network. The United States Congress should commit to funding existing and valued legacy programs such as the Land and Water Conservation Fund and enact new federal transportation legislation that supports the modern mobility and transportation interests, needs, and realities of American society. Financial support for greenway development is about return on investment. There are now dozens of economic impact studies proving that the return on investment in greenway development is substantial, at times ten dollars in return for every dollar invested. Dedicating a portion of federal transportation funding to greenway development has proven to increase transportation efficiency. America must commit the financial investment required to build out a nationwide greenway network. The return on investment will create jobs, grow businesses, and stimulate economic growth where it is needed the most.

System Mapping

Currently lacking are quality maps that clearly document and illustrate the magnitude of America's greenway and trail system. Quality mapping is attainable today using Geographic Information System technology. Needed are public and private sector partnering, cooperation, and investment to provide the data required to produce an accurate illustration of the size and extent of the network. The maps must illustrate planned and constructed facilities, the amount of conserved land that is part of the network, and gaps in the greenway network. Needed is an accurate portrayal of the size and magnitude of the current greenway network components. Having a picture of the current network that has developed over the past 100 years is fundamental for understanding where we are headed.

Branding, Marketing, and Promotion

The nationwide network of greenways needs to be adopted, promoted, and marketed across the nation and throughout the world as a significant achievement in which any nation would take great pride. Closing gaps in the network will be easier to complete through a coordinated branding, marketing, and promotion campaign. The resultant network will attract millions of visitors from around the world to experience the network and learn of its benefits. Business groups, such as the U.S. Chamber of Commerce, and federal agencies, such as the U.S. Travel and Tourism Advisory Board, in partnership with state, regional, and local equivalents, should be able to enthusiastically embrace and promote America's extensive network of completed greenways and trails. Critical information documenting the health, safety, educational, and financial benefits of travel and tourism should be disseminated across America. Most private sector business groups, along with public sector local, regional, state, and federal agencies, should be informed that travel and tourism promotes economic growth and is either the first, second, or third most important economic engine in all fifty states.

Closing Gaps

A concerted effort must be made among both public and private sectors to identify gaps (missing sections of greenway and trail in an otherwise continuous corridor) in the national greenway and trails network and implement a strategy for identifying, prioritizing, and closing those gaps. For example, one of the most significant issues to address and resolve is property ownership. Most gaps will involve working with private landowners to share documented benefits of a national greenway system.

Building the remaining sections of the system will take time and require the efforts of thousands of people; however, it is very much an achievable goal. This can be accomplished by challenging each state to adopt a vision and marshal the resources necessary to plan, design, and build cross-state greenway systems. For example, in North Carolina, the Mountains-to-Sea Trail is a more than 1,000-mile corridor extending east-west across the state, while the East Coast Greenway and Appalachian Trail provide north-south corridors. If each state could strive to achieve the same statewide trail and greenway vision that has occurred

Denotes approximate centerpoint
of gateway trails

The Great American Rail-Trail. Source: Rails-to-Trails Conservancy www.great-americanrailtrail.org.

in North Carolina, a national network of greenways would be realized. This has already occurred, to some extent, in many other states, including Florida, Michigan, Missouri, New York, and Pennsylvania.

Another approach is to develop a multistate partnership and program to build long-distance trails and greenways from coast to coast. One such project is already under way, the Great American Rail-Trail, envisioned by the Rails-to-Trails Conservancy (RTC) to span twelve states and the District of Columbia and approximately 4,000 miles in total length (eclipsing the length of the East Coast Greenway by 1,000 miles). The Great American Rail-Trail would, according to RTC, serve the needs of more than 50 million people (those who would live within fifty miles of the proposed trail route). The RTC has already identified what they refer to as "Gateway Trails," a dozen existing trails that serve as key connectors in the 4,000-mile route. These existing trails include the Capital Crescent Trail, Chesapeake and Ohio Canal National Historic Park, Panhandle Trail, Ohio to Erie Trail, Cardinal Greenway, Hennepin Canal Parkway, Cedar Valley Nature Trail, Cowboy Recreation and Nature Trail, Casper Rail-Trail, Headwaters Trail System, Trail of the Coeur d'Alenes, and the Palouse to Cascades State Park Trail. For more information about this exciting cross-country trail, visit www.greatamericanrailtrail.org.

The vision and programmatic activities of the Rails-to-Trails Conservancy are exactly the types of proposals and actions that are needed to build a network of coast-to-coast greenways that would transform America.

Why a National Greenway System Is Important

When the National Trail System Act was passed, it was with the clear intention "to promote the preservation of, public access to, travel within, and enjoyment and appreciation of the open-air, outdoor areas and historic resources of the Nation." Fifty years later there are several additional compelling reasons for completing and implementing a nationwide network of greenways and trails.

Establish a Conservation Ethic

America needs a conservation ethic now more than at any time in our nation's history, especially as the U.S. population continues to grow at accelerated rates. The resultant impacts to natural resources are increasingly more significant as more native landscapes and open space lands are converted to urban and suburban developments. Documented global climate changes are altering the native ecological systems of North America. Americans built the nation by exploiting natural resources, harvesting the bounty, and subsequently creating one of the richest nations the world has ever known. Aldo Leopold observed "We abuse land because we see it as a commodity belonging to us. When we see land as a community to which we belong, we may begin to use it with love and respect." Our land, water, and nature itself needs to be viewed as more than a commodity. Are we smart enough to recognize that conservation and economic prosperity go hand-in-hand?

The future of humankind has come into sharp focus as a result of global climate change and planet warming. A critically important fact not often discussed is the precious and precarious nature of all life on planet Earth. Our biosphere, where the majority of life exists, is an extremely thin layer of the planet that contains the key ingredients to support life. Thus far we know of no other zone of life, anywhere in the Universe, that is equal to our biosphere. While this fact is perhaps not well known by all, it is our reality. Can we continue to abuse, degrade, pollute, and otherwise denigrate this critically important ecological construct of our planet? Greenways alone are not the panacea, but they do offer proven strategies for those wanting to mobilize and be part of solutions to known problems.

We must continue the conservation work begun by Laurance Rockefeller in the 1960s. Building a national network of greenways would strengthen America's commitment to land and water conservation—reaffirming a conservation ethic. Greenways are replicable landscapes that can be developed at different sizes and scales but require the cooperation and commitment of a community of people in order to maximize their tangible benefits. Our nation and our world need a more balanced approach to resource conservation versus resource exploitation. Greenways embody the principles of a conservation ethic, which sustains the health of the natural world while at the same time providing access to these life-sustaining resources.

Greenways as "Gene-Ways"

Human activity is the root cause placing the world on the cusp of a mass extinction of plants and animals. As humans are the cause of this extinction, humans can also become the solution! A national network of greenways must be part of a comprehensive strategy that protects, conserves, and stewards the biological diversity of biomes and regional landscapes. One of the greatest benefits of a national greenway network may be in its ability to form critically important migratory "gene-way" corridors designated for plants and animals.[9]

Landscape architect Paul Cawood Hellmund in *The Ecology of Greenways* describes the framework for a "gene-way" as "an ecological imperative for anyone who has anything to do with the care of land, whether as visitor, owner, designer, or manager. Ethical treatment of natural processes, including the movement of plants, animals, and water, demands a right of passage for nature, preferably through areas that have historically served as conduits but at least through suitable adjacent areas."[10] Today, the consequences of global climate change and a warming planet need us to consider the "right of passage" of plants and animals to move about the landscape, without barrier and restriction, so they can adapt to changing conditions. Required are corridors dedicated to the purpose of wildlife and plant migration in response to climate change.

The United States must become more resilient to the impacts associated with a dynamic and changing climate as the merits of an interconnected framework of greenways and gene-ways are considered. The Fourth National Climate Assessment report, released in November 2018, assessed the science of climate change and its harmful impacts across the United States, offering a challenging view of the future: "Climate change creates new risks and exacerbates existing vulnerabilities in communities across the United States, presenting growing challenges to human health and safety, quality of life, and the rate of economic growth. Without substantial and sustained global mitigation and regional adaptation efforts, climate change is expected to cause growing losses to American infrastructure and property and impede the rate of economic growth over this century."

To build a resilient ecological framework that serves the needs of nature and civilization, a national interconnected network of greenways must be part of a mitigation and regional adaptation strategy across America. Greenways are among the most resilient landscapes, especially when they are appropriately designed and developed, to absorb the impacts resulting from extreme weather events. Are we as a nation willing and able to invest the resources required to complete a national greenway network as a mission critical resilient landscape framework? This involves ending the debate over the reality of climate change and taking actions necessary to adjust and adapt to its impacts. A national network of greenways that responds to ecological and environmental rather than human-defined jurisdictional boundaries is an important first step. This requires a coordinated approach between federal, state, and local governments, in partnership with diverse landowners and private sector interests.

Ensure Public Health, Safety, and Well Being

Greenways serve many useful and functional purposes as documented in the preceding chapters. A national network of greenways can provide safe passage for local travel during extreme weather events, vegetated buffers across large regional landscapes that serve as fire breaks in arid landscapes or absorption zones for excess water, critically important aquifer recharge zones where surface water is allowed to pool and slowly percolate through the soil, and opportunity for "forest bathing" (healing properties associated with walking through forested landscapes) and mental wellness, as relief from stressful urban living conditions.

In 2003, surgeon general Richard Carmona declared obesity a chronic disease and a national health epidemic, with nearly two of three Americans classified as being overweight or obese and one of every eight deaths attributed to obesity. As the surgeon general stated in his report to Congress, obesity is a preventable illness that can be effectively managed by encouraging Americans to become more active in their daily lives. Greenways provide affordable, close to home and work activity zones that encourage people to go outdoors and exercise each day as part of daily mobility and travel. A national network of greenways would provide local, regional, and statewide landscapes necessary to support active lifestyle communities across the nation. Imagine if each community in America were challenged to create a locally sponsored greenway that residents could use to support an active lifestyle. This would enable America to meet the challenge put forth by the U.S. surgeon general, and other leading health organizations, to develop these affordable Active Community Environments (ACES).[11] The United States has spent considerable time during the past two decades debating the merits of a national health care system. Are we prepared to invest in active community solutions, such as a national greenway network, as an affordable way to promote a healthier population? A national network of greenways can provide the affordable cure that America needs to effectively address a range of preventable illnesses.

Support Efficient Mobility and Transportation

One of the legacies of modern American greenways is their ability to serve as effective nonmotorized, nonpolluting travel corridors within local and regional transportation networks. Each year millions of Americans lose productive time stuck in peak travel automobile traffic jams. This loss of productivity registers in the calculus of the nation's Gross Domestic Product. Bicycling, walking, and transit are effective and viable local and regional transportation solutions, particularly when "trip-chaining" (combined methods of bicycling, walking, and transit riding as part of a single trip) occurs. Embracing and adopting a national greenway network as a critically important component of America's national transportation framework is important. Such adoption would elevate the status of bicycling and walking in line with other modes of travel, codify that bicycling and walking is an important transportation pursuit in all fifty states, and provide direction within states' departments of transportation.

The emergence of autonomous vehicles, driverless trucks, delivery drones, intelligent highways, metro and regional mass transit systems, and electric bicycles are strong indicators of a new era in transportation and mobility. A national network of greenways can and must be an integral component of this new paradigm of travel and transport. Will the United States enact a new federal transportation legislation that launches America into a new era of personal mobility and transportation that is not reliant on the automobile? Do we have the courage and foresight to understand that new methods and new modes of travel and transport are in the future of travel?

Celebrate Culture and Diversity

Greenways are democratic landscapes, available to all citizens without prejudice. A national network of greenways can connect all Americans to history and culture in a manner that celebrates our diversity. What has made the United States of America unique among all other nations of the world is the nation's diverse culture; we are a nation of immigrants who historically overcame cultural differences to collectively live in a democratic society and pursue economic opportunity. American society and its opportunities have not been uniformly available to all, as the cloud of racism has been perpetuated throughout the nation's history. Are we as a nation smart enough to overcome our cultural failures and wise enough to embrace the richness of our diversity?

Greenway planners, designers, and builders need to listen to the concerns of neighborhood residents and fit greenway design, development, and operating programs to suit local needs and interests. This is a different approach from taking a greenway program and trying to fit it into the neighborhood. Greenway enthusiasts must recognize that a community places higher priority on resolving crime and improving public safety, building more affordable housing, improving access to health care, creating new well-paid jobs, fostering better education, and improving access to quality food. The question to ask is how would construction of a greenway address the most pressing needs of community residents?

America should expand its cultural and historic trails as part of building out the national greenway network. Some historic routes already exist, such as the Trail of Tears, the Underground Railroad, and the Santa Fe Trail. Metropolitan areas are also beginning to build a diverse array of cultural trails, with the Indianapolis Cultural Trail perhaps being one of the better-known examples.

Expand the Experience Economy

An interconnected network of greenways across America can provide a seamless travel experience, transporting people from urban centers to rural communities. A national network of greenways and trails will enable Americans and foreign visitors to pursue unique outdoor experiences for cultural enrichment, physical challenge, access to unique native landscapes, and adventure travel. The world economy is in the midst of an important economic revolution in which experiences become more valuable to consumers than the instruments they purchased to make the experience possible (for example, plane ticket, hotel, bicycle, backpack, rental car). "Economists have typically lumped experiences in with services, but experiences are a distinct economic offering, as different from services as services are from goods. Today we can identify and describe this fourth economic offering because consumers unquestionably desire experiences, and more and more businesses are responding by explicitly designing and promoting them."[12]

The Ownership Economy is fading, and the Sharing Economy is growing. American consumers are increasingly interested in purchasing unique and memorable experiences. The most valued companies in the world today like, Apple, Facebook, and Google, are those that drive the Experience Economy. These highly valued companies don't just produce products that enrich life; they seek to enhance life's experiences through their increasingly diverse consumer services. Facebook, for example, enables one to share experiences, as they occur, with family members and friends. Google can be used to model travel preferences and match one's desires with companies that provide the most appropriate services to make those experiences as rewarding as possible.

A national greenway network has the potential to offer a diverse array of experiences that would serve to expand the American economy. Travel and tourism are either the first, second, or third most important economic engine in all fifty states. The Experience Economy is growing faster than other segments of the American economy, as substantiated by the growth of the outdoor industry. Travel, tourism, and an active lifestyle will fuel America's economy in the decades to come. Does America have the financial wherewithal to realize the substantial return on investment possibly derived from investing in a national network of interconnected greenways?

Conclusion

The greenway stories in this book are a small and hopefully representative sample of how people have come to embrace the concept of greenways across America and around the world. There are many more greenway stories to be written.

The building of greenways transcends the technical narrative that is so critical to successful development. Once a greenway has been planned, designed, constructed, and opened for public use and enjoyment, very few people realize or remember the scientific assessment, design, engineering, and landscape architectural detailing that made the project possible and successful. What the vast majority of greenway users will know is that they have been connected to their friends, work, school, and other places by a special landscape. Building a greenway fulfills the desire to be connected to resources within a community that people know and want to save for others to experience. This is why so many Americans have become champions of local greenways.

Connections are the underlying imperative that motivated Manning to author a "National Plan" for America and which became the fundamental underpinning of the President's Commission on Americans Outdoors signature recommendation for a national network of greenways. During the past hundred years America has transformed Manning's 1919 vision for a national interconnected trail and greenway network into reality. The next hundred years will realize even more progress. In the not so distant future Americans will be able to traverse our nation on a dedicated network of north-south and east-west off-road trails and greenways. This network will provide the fuel to grow the economy, provide critical migratory routes for plants and animals, enrich communities large and small, and link Americans to our heritage—and to landscapes worthy of exploration and enjoyment.

Wolf River Greenway bridge installation. Source: Wagner Construction Company.

ACKNOWLEDGMENTS

I want to acknowledge and thank my family, friends, and colleagues who helped me transform this book from vision to reality. My wife, Marjorie, patiently worked with me through the process of composing each chapter, providing valuable feedback and editorial guidance. My mother, Jane, and sister, Jeni, helped me frame the story format during the early stages of this book. I am thankful that my dad was able to read a couple of the chapters before he passed away in 2012.

A very special thanks to my friend, colleague, and mentor Professor Gene Bressler at North Carolina State University, who knew I needed a kick in the pants to finish the manuscript. Gene was also instrumental in helping me focus and hone the manuscript.

Thank you to Keith Laughlin for authoring a terrific foreword for this book. Thank you to Alana Brasier and Paul Cawood Hellmund for reviewing a not-ready-for-prime-time manuscript. Your feedback was immensely helpful. A huge thank you to Tim Patterson, Andy Fox, Rodney Swink, Mike Greene, Dan Howe, Robby Layton, Brenda McClymonds, Bill Eubanks, Sig Hutchinson, Daniel Mourek, and Lisa Watts for taking the time to complete individual chapter reviews and submit constructive feedback.

I am extremely grateful to Glenn Morris for connecting me with his sister (my publisher) and Michael Savage for his wise and timely counsel. Thanks to Jason Reyes, Laura England, Frank McGuire, Davitt Woodwell, Dennis Markatos-Soriano, Edwin Torres, Jeff Hawkins, Karen Haley, Natalie Lozano, Outside Las Vegas, the Trust for Public Land, the Rails-to-Trails Conservancy, and the Grand Forks Visitor and Convention Bureau for giving me permission to use photos and graphics that help describe the richness of each project.

A sincere and heartfelt thank you to the staff of Greenways Incorporated and Alta Planning + Design, who worked with me over the course of 35-plus years to plan, design, and oversee the construction of the projects featured in this book.

Thank you to Meredith Morris-Babb for understanding the potential of the manuscript, and to the entire team at the University of Florida Press for polishing and presenting the work. Project editor Marthe Walters, copy editor Gillian Hills, and director of marketing and sales Romi Gutierrez have been fantastic to work with during the past two years.

Finally, this book would never have been possible without the vision, aspirations, and efforts of my clients and colleagues. My professional career, and indeed my life, is so much richer for having had the opportunity to work alongside each of you to address and resolve the unique problems and issues of your community and landscape. Hopefully, I have accurately and faithfully depicted the significant contributions that each of you has made in advancing the American greenway movement. Best wishes and happy trails.

GLOSSARY

biome: "A community of plants and animals that have common characteristics for the environment they exist in. They can be found over a range of continents. Biomes are distinct biological communities that have formed in response to a shared physical climate. Biome is a broader term than habitat; a biome can comprise a variety of habitats."[1]

floodplain: An area of land adjacent to a stream or river which stretches from the banks of its channel to the base of the enclosing valley walls, and which experiences flooding during periods of high discharge.[2]

floodway: A "Regulatory Floodway" means the channel of a river or other watercourse and the adjacent land areas that must be reserved in order to discharge the base flood without cumulatively increasing the water surface elevation more than a designated height. Communities must regulate development in these floodways to ensure that there are no increases in upstream flood elevations. For streams and other watercourses where FEMA has provided Base Flood Elevations (BFEs), but no floodway has been designated, the community must review floodplain development on a case-by-case basis to ensure that increases in water surface elevations do not occur, or identify the need to adopt a floodway if adequate information is available.[3]

gentrification: A process of renovating a deteriorated urban neighborhood by means of influx of more affluent residents. This is a common and controversial topic in politics and in urban planning. Gentrification can improve the material quality of a neighborhood, while initially forcing relocation of current, established residents and businesses, causing them to move from gentrified areas, seeking lower-cost housing.[4]

green infrastructure: "Used as a noun, green infrastructure refers to an interconnected green space network (including natural areas and features, public and private conservation lands, working lands with conservation values, and other protected open spaces) that is planned and managed for its natural resource values and for the associated benefits it confers to human populations. Used as an adjective, green infrastructure describes a process that promotes a systemic and strategic approach to land conservation at the national, state, regional and local scales, encouraging land-use planning and practices that are good for nature and for people."[5]

"Green infrastructure is a cost-effective, resilient approach to managing wet weather impacts that provides many community benefits. While single-purpose gray stormwater infrastructure—conventional piped drainage and water treatment systems—is designed to move urban stormwater away from the built environment, green infrastructure reduces and treats stormwater at its source while delivering environmental, social, and economic benefits."[6]

greenprint: "A greenprint is a strategic conservation plan that recognizes the economic and social benefits that parks, open space, and working lands provide communities. Such benefits include recreation opportunities through the use of parks and trails, habitat protection and connectivity, clean water, agricultural land preservation, and increased resilience to climate change. Through the development of a greenprint, stakeholders help to identify, map, and prioritize areas important to the conservation of plants and wildlife, water resources, recreational opportunities, and working landscapes. A greenprint reflects local shared priorities and culture. Greenprints can be created largely through technical and scientific input, though they usually involve engagement from the general public and local conservation groups or local government.

In its initial form, a greenprint can be a map-based representation of the open space assets with natural resource and community-based values across the region. The map and associated data can help landowners, local governments, land trusts, and public agencies focus development away from important natural areas and working lands, prioritize conservation areas, and help the public understand the tradeoffs of various land use decisions."[7]

greenspace: greenspace or green space refers to protected areas of undeveloped landscape, land used for parks, native open areas or landscapes, or community space consisting of land without buildings that is covered by native vegetation.[8]

greenway: A corridor of undeveloped land preserved for recreational use or environmental protection.[9] "Greenways often follow natural features such as ridgelines or river valleys with features such as canals, utility corridors, abandoned rail lines, roadways or wherever a break appears in the land pattern. Although each greenway is unique, most greenways are networks of natural open space corridors that connect neighborhoods, parks and schools to areas of natural, cultural, recreational, scenic and historic significance. These passageways link people and places to nature for the enjoyment and enrichment of the community. They can be classified into regional, city and community greenways."[10]

landscape architecture: The design of outdoor areas, landmarks, and structures to achieve environmental, social-behavioral, or aesthetic outcomes. It involves the systematic investigation of existing social, ecological, and soil conditions and processes in the landscape, and the design of interventions that will produce the desired outcome. The scope of the profession includes landscape design, site planning, stormwater management, erosion control, environmental restoration, parks and recreation planning, visual resource management, green infrastructure planning, private estate and residence landscape master planning and design, all at varying scales of design, planning, and management.[11]

landscape typology: The classification and description of natural and cultural landscapes. Generally, landscapes can be defined and classified by their dominant natural features and the condition of disruption or change by natural forces, or influence of human activity (e.g., urbanization, agriculture).[12]

megaregion: A large network of metropolitan regions that share several or all of the following characteristics: environmental systems and topography, infrastructure systems, economic linkages, settlement and land use patterns and culture and history. As of 2018, more than 70 percent of the U.S. population and jobs are located in eleven megaregions. Megaregions are becoming new competitive units in the global economy, characterized by the increasing movement of goods, people, and capital among their regions.[13]

recyclables: Objects that are able to be processed and used again.[14]

resiliency: The power or ability to return to the original form, position, etc. after being bent, compressed, or stretched; elasticity. Ability to recover readily from illness, depression, adversity, or the like; buoyancy.[15] "Urban resilience is the capacity of individuals, communities, institutions, businesses, and systems within a city to survive, adapt, and grow no matter what kinds of chronic stresses and acute shocks they experience."[16]

sustainability: The ability to be maintained at a certain rate or level. Avoidance of the depletion of natural resources in order to maintain an ecological balance.[17]

"Sustainability is the process of maintaining change in a balanced fashion, in which the exploitation of resources, the direction of investments, the orientation of technological development and institutional change are all in harmony and enhance both current and future potential to meet human needs and aspirations."[18]

NOTES AND SOURCES

Introduction

1. Charles Little, *Greenways for America* (Washington, D.C.: Island Press, 1990).
2. Ibid., 203.

Chapter 1. A Close Family Legacy: Anne Springs Close Greenway, Fort Mill, South Carolina

1. Francis Bradley, "Palmetto Place Names," *The State* (Columbia, S.C.), 1958.
2. Elliott White Springs, *War Birds: The Diary of a Great War Pilot* (Havertown, UK: Frontline Books, 2016). Originally published 1927.

ADDITIONAL SOURCES

Allen, Debra J., et al. South Carolina Department of Historic Resources, National Register of Historic Places Registration for *William Elliott White House*, February 1987, United States Department of the Interior.

Gettys, Paul. Catawba Regional Planning Council, National Register of Historic Places Registration for *Nation's Ford Road*, August 1997, United States Department of the Interior.

———. Catawba Regional Planning Council, National Register of Historic Places Registration for *Springfield House and Plantation*, August 1985, United States Department of the Interior.

Greenways Incorporated. *100-Year Vision, 25-Year Master Plan, 5-Year Action Plan for the Anne Springs Close Greenway*. Leroy Springs and Company, 1992.

Leroy Springs and Company. "Anne Springs Close Greenway." Retrieved from https://www.ascgreenway.org.

———. Anne Springs Close Greenway Facebook page. Retrieved from https://www.facebook.com/ASCGreenway/.

McHarg, Ian. *Design with Nature*. New York: Wiley, 1995. Originally published 1969.

Morse, LeAnne Burnett. *Images of America, Fort Mill*. Charleston, S.C.: Arcadia Publishing, 2015.

South Carolina Encyclopedia. "Treaty of Augusta, November 1763." Retrieved from http://www.scencyclopedia.org/sce/entries/treaty-of-augusta/, and the original treaty retrieved from http://vault.georgiaarchives.org/cdm/ref/collection/adhoc/id/1590.

Chapter 2. Come Hell and High Water: Greater Grand Forks Greenway, Grand Forks, North Dakota

1. Liz Creel, "Ripple Effects: Population and Coastal Regions" (Washington, D.C.: Population Reference Bureau, 2003).

2. National Oceanic and Atmospheric Administration, *National Coastal Population Report: Population Trends from 1970 to 2020*, March 2013.

ADDITIONAL SOURCES

Arvidson, Adam Regn. "When Rivers Rise." *Landscape Architecture* 100, no. 2 (February 2010): 78–90.

Flink, Charles, et al. Greater Grand Forks Greenway Master Plan. City of Grand Forks, North Dakota, September 2001.

Greendahl, Kim, and Maegin Rude. The Greenway 2012 Supplement. City of Grand Forks, North Dakota, 2012

Karlen, Neal. "36 Hours in Grand Forks." *New York Times*, February 10, 2006.

Maidenberg, Mike, Mike Jacobs, and Tom Dennis. Various articles in the *Grand Forks Herald* newspaper, 1999–2001.

Steiner, Christopher. "River Revival." *Fortune Small Business*. May 2006.

"The Greenway." Retrieved from http://www.greenwayggf.com.

Chapter 3. Turning Trash into Trails: Swift Creek Recycled Greenway, Cary, North Carolina

1. Paul Hawken, Amory Lovins, and L. Hunter Lovins, *Natural Capitalism* (Boston: Little, Brown and Co., 1999).

2. Ibid.

3. Brundtland Commission, *Our Common Future: Report of the World Commission on Environment and Development* (Oxford, U.K.: Oxford University Press, 1987).

4. U.S. Energy Information Administration. Retrieved from www.eia.gov.

5. U.S. Environmental Protection Agency, "Facts and Figures about Materials, Waste and Recycling." Retrieved from https://www.epa.gov.

6. U.S. Environmental Protection Agency. Retrieved from www.archive.epa.gov.

7. Amount of waste recycled in 2008.

8. U.S. Environmental Protection Agency, "Facts and Figures about Materials, Waste and Recycling." Retrieved from https://www.epa.gov.

9. Aluminum Association, "The Aluminum Can Advantage: Key Sustainability Performance Indicators for the Aluminum Can." Retrieved from https://www.aluminum.org/aluminum-can-advantage.

10. Adam Millard-Ball, "Where the Action Is: A Sustainable Places Story," *Planning* 76, no. 7 (August/September 2010): 16–21.

11. Cody Marshall, and Karen Bandhauer, "The Heavy Toll of Contamination," *Recycling Today* April 19, 2017. Retrieved from recyclingtoday.com: http://www.recyclingtoday.com/article/the-heavy-toll-of-contamination/.

12. Nick Stockton, "Listen Up America: You Need to Learn How to Recycle. Again," *Wired* August 21, 2015. Retrieved from wired.com: https://www.wired.com/2015/08/listen-america-need-learn-recycle/.

13. Will Flower, "What Operation Green Fence Has Meant for Recycling," *Waste 360* February 11, 2016. Retrieved from waste360.com: http://www.waste360.com/business/what-operation-green-fence-has-meant-recycling.

ADDITIONAL SOURCES

Braungart, Michael, and William McDonough. *Cradle to Cradle: Remaking the Way We Make Things*. New York: North Point Press, 2002.

Town of Cary, North Carolina. "Swift Creek Greenway." Retrieved from: https://www.townofcary.org/recreation-enjoyment/parks-greenways-environment/greenways/swift-creek-greenway.

Chapter 4. Something Grand: Grand Canyon Greenway, Grand Canyon National Park, Arizona

1. Address of President Roosevelt at Grand Canyon, Arizona, May 6, 1903. Theodore Roosevelt Papers. Library of Congress Manuscript Division. https://www.theodorerooseveltcenter.org/Research/Digital-Library/Record?libID=0289796. Theodore Roosevelt Digital Library. Dickinson State University.

2. Robert McNamara, "Artist George Catlin Proposed Creation of National Parks," *ThoughtCo.* August 1, 2019. Retrieved from https://www.thoughtco.com/proposed-creation-of-national-parks-1773620.

3. Alfred Runte, *National Parks: The American Experience*, 4th ed. (Lanham, Md.: Taylor Trade Publishing, 2010).

4. Rolf Diamant, "The Olmsteds and the Development of the National Park System," National Association for Olmsted Parks, 2008. Retrieved from: http://www.olmsted.org/the-olmsted-legacy/the-olmsted-firm/the-olmsteds-and-the-development-of-the-national-park-system.

5. UNESCO World Heritage designation for Grand Canyon National Park. Retrieved from https://whc.unesco.org/en/list/75.

6. National Park Service, *General Management Plan for Grand Canyon National Park*, Shapins Associates, 1995. Retrieved from https://parkplanning.nps.gov/ManagementPlans.cfm.

7. Jeff Olson, *The Third Mode: Towards a Green Society*. Self-published, 2012.

8. First Lady Hillary Rodham Clinton, speech at the dedication ceremony of the Grand Canyon Greenway, April 1999.

ADDITIONAL SOURCES

Flink, Charles, et al. *Grand Canyon Greenway Master Plan*. The Greenway Collaborative, July 1997.

Chapter 5. Open Space in Vegas—It's a Sure Bet: Las Vegas Open Space and Trails, Las Vegas, Nevada

1. Mark A. Benedict and Edward T. McMahon, *Green Infrastructure: Linking Landscapes and Communities* (Washington, D.C.: Island Press, 2006).

2. The goal was to conduct a statistically valid survey, based on an accurate and controlled survey of the resident population. ETC set a target of collecting 500 completed surveys from households throughout northwest Las Vegas. The actual survey yielded 688

random household responses, with a 95 percent confidence factor and a margin of error ± 3.7 percent.

ADDITIONAL SOURCES

Get Outdoors Nevada. *Vegas Valley Rim Trail Plan*. 2013.

Greenways Incorporated et al. *Las Vegas Trails Gap Analysis Study*. City of Las Vegas, April 2005.

———. *Northwest Las Vegas Open Space Plan*. City of Las Vegas, January 2005.

———. *SNRPC Regional Open Space Plan*. Southern Nevada Regional Planning Coalition, May 2006.

———. *Floyd Lamb Park Master Plan*. City of Las Vegas, April 2007.

Planning Department, City of Las Vegas. *Vegas Master Plan 2020*. City of Las Vegas, May 2012.

Southern Nevada Health District. *Neon to Nature: Get Healthy Clark County*. Retrieved from https://gethealthyclarkcounty.org/get-moving/community-activities/neon-to-nature/.

Chapter 6. Miami Means "Sweet Water": Miami River Greenway, Miami, Florida

1. "Miami Circle," Wikipedia, last updated 6 June 2019, 11:05, https://en.wikipedia.org/wiki/Miami_Circle.

2. "Miami River (Florida)," Wikipedia, last updated 2 June 2019, 22:56, https://en.wikipedia.org/wiki/Miami_River_(Florida).

3. Miami River Commission, "Miami River: A Working River," 2008. Retrieved from www.miamirivercommission.org.

4. Miami River Commission Annual Report, 2017. Retrieved from http://www.miamirivercommission.org/PDF/MRC2017ReportWeb.pdf.

5. Peter Layne Taylor, "Why Are the World's Top Real Estate Investors Risking Billions on Miami's Riverfront Renaissance?" *Forbes* December 9, 2016.

ADDITIONAL SOURCES

Benfield, Kaid. "The Death and Life of Little Havana." *City Lab* September 14, 2012. https://www.citylab.com/equity/2012/09/death-and-life-little-havana/3275/.

Beyer, Scott. "How Miami Fought Gentrification and Won (For Now)." *Governing* July 2015. https://www.governing.com/columns/urban-notebook/gov-miami-gentrification.html

Friends of the Miami River Greenway Facebook page. Retrieved from https://www.facebook.com/MiamiRiverGreenway/.

Greenways Incorporated. Miami River Greenway Plan. April 2001. Retrieved from www.storiesfromthetrail/miamiriver.pdf.

"Malecón, Havana 2018." Wikipedia. Last updated 14 July 2019, 06:05. https://en.wikipedia.org/wiki/Malecón,_Havana.

Miami-Dade County. Miami-Dade Greenprint. 2015. Retrieved from https://www.miamidade.gov/greenprint/pdf/climate_action_plan.pdf.

Smiley, David. "Upzoning of East Little Havana Scrapped as Miami Planners Go Back to the Drawing Board." *Miami Herald* January 24, 2017.

The Trust for Public Land. "Miami River Greenway." Retrieved from https://www.tpl.org/our-work/miami-river-greenway#sm.00014aq0dofdxf9ipi718hd02atmq.

Chapter 7. Lowcountry Life: Charleston County Greenbelt Plan, Charleston County, South Carolina

1. Ross W. Gorte, *U.S. Tree Planting for Carbon Sequestration*, Congressional Research Service, May 4, 2009, https://fas.org/sgp/crs/misc/R40562.pdf.
2. U.S. Forest Service Carbon Online Estimator.

ADDITIONAL SOURCES

City of Charleston. "Charleston Green Plan: A Roadmap to Sustainability." 2009. https://www.charleston-sc.gov/DocumentCenter/View/1458/Charleston-Green-Plan?bidId=.
Flink, Charles, Jason Reyes, Cathy Ruff, et al. Charleston County Greenbelt Plan. June 2006.
Hiss, Tony. *The Experience of Place*. New York: Vintage Books, Random House, 1991.

Chapter 8. Callin' the Hogs: The Northwest Arkansas Razorback Regional Greenway, Arkansas

1. Walton Family Foundation, "Walton Family Foundation: Where Opportunity Takes Root." Retrieved from https://www.waltonfamilyfoundation.org/about-us.
2. Susan Handy et al., "The Regional Response to Federal Funding for Bicycle and Pedestrian Projects," Institute of Transportation Studies, University of California, Davis, July 1, 2009.
3. Robert Puentes, "Transportation Reform of 1991 Remains Relevant," Brookings Institution, December 19, 2011, https://www.brookings.edu/blog/up-front/2011/12/19/transportation-reform-of-1991-remains-relevant/.
4. I arrived at this number by adding up trails built between 1991 and 2018, using a variety of sources.
5. Walton Family Foundation, Northwest Arkansas Trail Usage Monitoring Report, 2017.
6. Jeremy Pate, "Blazing a New Trail for Bike Commuters in Northwest Arkansas," Walton Family Foundation, November 15, 2017, https://www.waltonfamilyfoundation.org/stories/home-region/blazing-a-new-trail-for-bike-commuters-in-northwest-arkansas.
7. "Economic and Health Benefits of Bicycling in Northwest Arkansas," prepared for The Walton Family Foundation and PeopleForBikes (Denver, Colo.: BBC Research and Consulting, 2018).
8. Charles Little, Greenways for America (Washington, D.C.: Island Press, 1990).

ADDITIONAL SOURCES

Northwest Arkansas Razorback Regional Greenway TIGER II Grant Application. Alta Planning + Design and the Northwest Arkansas Regional Planning Commission. August 2010.

Chapter 9. White Russia: International Greenway Efforts in Belarus

1. Charles Flink, *Greenways as Agrotourism Corridors*, Country Escape Belarus and the Embassy of the United States, September 2012.

2. Ibid.

3. Civic Tourism, "Rethink Economics," 2010, http://www.civictourism.org/strategies1.html.

4. Economic Development Partnership of North Carolina, "North Carolina Tourism Supports Record Employment and Visitor Spending in 2017," May 28, 2018, https://edpnc.com/north-carolina-tourism-supports-record-employment-visitor-spending-2017/.

ADDITIONAL SOURCES

Country Escape. "Greenways in Belarus." http://www.greenways.by/index.php?lang =en.

European Greenway Association. "About the EGWA." http://www.aevv-egwa.org.

Friends of Czech Greenways. "The Prague-Vienna Greenways." http://www.praguevien-nagreenways.org.

"Naliboki Forest, Belarus." Wikipedia. Last updated 13 December 2018, 19:27. https:// en.wikipedia.org/wiki/Naliboki_forest.

Outdoor Foundation. "Outdoor Recreation Participation Topline Report 2017." https:// outdoorindustry.org/wp-content/uploads/2017/04/2017-Topline-Report_FINAL. pdf.

United States Department of State. Programs and Events, Embassy of the United States Minsk Belarus, 2012. Retrieved from http://minsk.usembassy.gov/flink.html.

Chapter 10. America's Longest Urban Greenway: East Coast Greenway, from Maine to Florida

1. East Coast Greenway Alliance, 2018 usage figures. Retrieved from https://www.greenway.org.

2. Alta Planning + Design, "The Impact of Greenways in the Triangle: How the East Coast Greenway Benefits Health and Economy of North Carolina's Triangle Region," September 2017, https://altaplanning.com/wp-content/uploads/Health-and-Economic-Benefits-of-East-Coast-Greenway-to-North-Carolina%E2%80%99s-Triangle-Region.pdf.

3. President's Commission on Americans Outdoors, *Americans Outdoors: The Legacy, the Challenge, with Case Studies, the Report of the President's Commission* (Washington, D.C.: Island Press, 1987).

4. Joseph Pine and James Gilmore, *The Experience Economy: Work Is Theater and Every Business a Stage* (Boston: Harvard Business Review Press, 2011).

ADDITIONAL SOURCES

Bush White House, Anne Lusk, VT, Greenways Incorporated, and Charles A. Flink. Papers. 1970–2014, MC 00405, Box 62, Folder 4, North Carolina State University Special Collections Research Center, Raleigh, North Carolina.

East Coast Greenway Alliance. State of the Trail Report. Karen Votava, editor. Wakefield, Rhode Island, January 2001.

———. History of the East Coast Greenway, 1991–2009. Karen Votava, editor, Wakefield, Rhode Island. January 2009.

———. Annual Report. Dennis Markatos-Soriano. Durham, North Carolina. January 2015.

———. Annual Report. Dennis Markatos-Soriano. Durham, North Carolina. January 2017.

Chapter 11. A National Greenway System: Envisioning a Coast-to-Coast Greenway System

1. Warren Manning, "A National Plan," 1. 1919. Unpublished manuscript. Retrieved from Iowa State University Library Digital Collections: https://digitalcollections.lib.iastate.edu/warren-h-manning.

2. Ibid., 205.

3. Ibid., 245.

4. National Park Service, "National Trails System Act Legislations," https://www.nps.gov/subjects/nationaltrailssystem/national-trails-system-act-legislation.htm.

5. President's Commission on Americans Outdoors, "Report and Recommendations to the President of the United States," (Washington, D.C., 1986).

6. George H. Siehl, "The Policy Path to the Great Outdoors: A History of the Outdoor Recreation Review Commissions," working paper, Resources for the Future (www.rff.org), 2008.

7. President's Commission on Americans Outdoors, "Report and Recommendations to the President of the United States," (Washington, D.C., 1986).

8. Rails-to-Trails Conservancy, Washington, D.C., http://www.railstotrails.org/our-work/united-states/, and American Hiking Society, *Hiking Trails in America: Pathways to Prosperity* (Washington, D.C., June 2015).

9. What is meant by the word "gene-way?" It's a play on the word greenway, rather than a literal interpretation, with the emphasis on conserved and protected corridors of land and water that accommodate the biological needs of plants and animals.

10. Paul Cawood Hellmund and Daniel S. Smith, *Ecology of Greenways: Design and Function of Linear Conservation Areas* (Minneapolis: University of Minnesota Press, 1993), 210.

11. US Department of Health and Human Services, Centers for Disease Control and Prevention, National Center for Chronic Disease Prevention and Health Promotion, *Active Community Environments Brochure*, June 2000.

12. Joseph Pine and James Gilmore, *The Experience Economy: Work Is Theater and Every Business a Stage* (Boston: Harvard Business Review Press, 2011).

ADDITIONAL SOURCES

Ahern, Jack. *Greenways as Strategic Landscape Planning: Theory and Application.* The Netherlands: Wagenigen University, 2002.

Fabos, Julius, and Jack Ahern, eds. *Greenways: The Beginning of an International Movement.* New York: Elsevier, 1995.

Flink, Charles, et al. *Greenways: A Guide to Planning, Design and Development.* Washington, D.C.: Island Press, 1993.

Little, Charles. *Greenways for America*. Baltimore, Md.: Johns Hopkins University Press, 1990.

Glossary

1. "Biome," Wikipedia, last updated 19 July 2019, 22:39, https://en.wikipedia.org/wiki/Biome.

2. "Floodplain," Wikipedia, last updated 26 July 2019, 13:40, https://en.wikipedia.org/wiki/Floodplain.

3. FEMA.gov.

4. "Gentrification," Wikipedia, last updated 4 August 2019, 10:36, https://en.wikipedia.org/wiki/Gentrification.

5. Mark Benedict and Edward McMahon, *Green Infrastructure: Linking Landscapes and Communities* (Washington, D.C.: Island Press, 2006). Retrieved from www.epa.gov, 2018.

6. United States Environmental Protection Agency, "What Is Green Infrastructure?" last updated 29 May 2019, https://www.epa.gov/green-infrastructure/what-green-infrastructure.

7. The Nature Conservancy, "What Is a Greenprint?" Retrieved 2019, https://www.conservationgateway.org/ConservationPractices/PeopleConservation/greenprints/Pages/what.aspx.

8. Multiple sources.

9. Merriam-Webster.com, last updated 6 Aug 2019.

10. Charles Little, *Greenways for America* (Washington, D.C.: Island Press, 1990).

11. "Landscape architecture," Wikipedia, last updated 7 August 2019, 01:16, https://en.wikipedia.org/wiki/Landscape_architecture.

12. Multiple sources.

13. Regional Plan Association; "Megaregions of the United States," Wikipedia, last updated 4 August 2019, 11:25, from https://en.wikipedia.org/wiki/Megaregions_of_the_United_States.

14. *Collins English Dictionary*.

15. Paraphrased from Merriam-Webster.com.

16. Rockefeller Foundation, "100 Resilient Cities," October 2018, https://www.rockefellerfoundation.org/our-work/initiatives/100-resilient-cities/.

17. Paraphrased from Merriam-Webster.com.

18. "Sustainability," Wikipedia, last updated 8 August 2019, 10:02, https://en.wikipedia.org/wiki/Sustainability.

INDEX

CHARLES A. "CHUCK" FLINK, FASLA, is an award-winning author, planner, and landscape architect. Flink is founder and president of Greenways Incorporated and a graduate of North Carolina State University College of Design. He is widely regarded as a leading national and international greenway planner and designer, having completed project work for more than 250 communities within thirty-six states. Flink is a Fellow in the American Society of Landscape Architects. In 2001, Flink received a Merit Award from the American Society of Landscape Architects for his involvement and leadership with the Grand Canyon Greenway project. In 2019, Flink received the Watauga Medal from his alma mater.